CIVILIZING CYBERSPACE

THE FIGHT FOR DIGITAL DEMOCRACY

CIVILIZING CYBERSPACE

THE FIGHT FOR DIGITAL DEMOCRACY

KENNETH JAMES HAMER-HODGES

Studio of Books LLC
5900 Balcones Drive Suite 100
Austin, Texas 78731
www.studioofbooks.org
Hotline: (254) 800-1183

Ordering Information:
Special discounts are available on quantity purchases by corporations, associations, and others. For details, contact the publisher at the address above.

Printed in the United States of America.

ISBN-13: Softcover   978-1-964148-66-3
              eBook   978-1-964148-67-0

Library of Congress Control Number: 2024910986

What others are saying:

'This book is a must-read for anyone concerned with software quality, hacking, malware, and cyber-crime. It covers the waterfront on Industrial Strength Computer Science and faithful, trusted software.' - Capers Jones III is an author and expert on software engineering, software quality, function point cost estimation and an advisor on software quality to the governments of South Korea and Malaysia.

'In this provocative book, Ken argues that the monolithic general-purpose computer is outdated, unchanged architecturally since batch processing mainframes were locked in secure computer rooms. The tremendous technological advances since the Cold War mean that computer software permeates every facet of society, networked into complex global interactions under relentless attack. Ken argues these systems are vulnerable to bad actors using intelligent malware. These global shortcomings are destroying cyber-society. Ken's book is a fascinating insight into the enormous cyber-security problems of a 21st-century democratic society.' - Peter Venton - OBE, BSc, CEng, MIET - former Major Program Review Chairman for HMG UK.

ADA'S ANTHEM FOR COMPUTER SCIENCE

ʻ There who view mathematical science, not merely as a vast body of abstract and immutable truths, whose intrinsic beauty, symmetry and logical completeness, when regarded in their connexion together as a whole, entitle them to a prominent place in the interest of all profound and logical minds, but as possessing a yet deeper interest for the human race, when it is remembered that this science constitutes the language through which alone we can adequately express the great facts of the natural world, and those unceasing changes of mutual relationship which, visibly or invisibly, consciously or unconsciously to our immediate physical perceptions, are interminably going on in the agencies of the creation we live amidst: those who thus think on mathematical truth as the instrument through which the weak mind of man can most effectually read his Creator's works, will regard with especial interest all that can tend to facilitate the translation of its principles into explicit practical forms.'

A.A.L. 1843

DEDICATION

To Charles Babbage KH FRS (1791 – 1871), the English polymath, inventor, and mechanical engineer. His work demonstrated the *Infallible Automation* of computers as a science.

To A.A.L. (Ada Augusta, Countess of Lovelace, 1815-1852) for her translation from French, 'Sketch of the Analytical Engine, Invented by Charles Babbage' from the Bibliothèque Universelle de Genève, October 1842, No. 82 and her 'Notes' upon the memoir.

To Luigi F. Menabrea (1809-1896), of Turin, Officer of the Military Engineers, first Prime Minister of a united Italy, who earlier documented Babbage's discovery of the Analytical Engine and 'Infallible Automation.'

To Alonzo Church (1903-1995) and his student Alan Turing (1912-1954), they discovered two sides to computer science known as the Church-Turing Thesis.

To John "Jack" Michael Cotton (1931-2018) for everything he taught me. To Alan Lawrence (1941-2019), Peter Venton and other colleagues for their inspirational friendship in inventing the PP2501[1] at Plessey Telecommunications Research in the UK during the 1970s[2]. Together, we pioneered Industrial Strength Computer Science.

---

1   Also known as the Plessey System 250

2   Information Security and Data Protection, Object Lessons from PP250. https://slideplayer.com/slide/7947797/

To Industrial Strength Computer Science, the solution to cyber dictators, cybercrime, hacking, and the breakout of super- intelligent malware as a citizen's platform for a democratic cyber society.

To the world, enhanced by Artificial Intelligence and secured by Church-Turing machines that democratically protect all future generations.

To breeds of Church-Turing Machines as Trusted Commercial Computers that level the playing field to evolve a stable electronic future with trusted software. A world that sustains governments of the people, by the people and for the people.

Most of all, to my beautiful wife, Christine. Her constant encouragement, her skills, and her enthusiastic feedback were vital.

ABOUT THE AUTHOR

I grew up in Portsmouth, England, a well-defended naval town since Henry VIII was King. During post-WWII reconstruction, I attended Portsmouth Grammar School, then worked my way through the local community college studying electrical engineering and as an intern at work with emerging semiconductor technologies for computer memory systems. When I graduated in 1967, I had a job at hand. As a young engineer, I helped design memory systems for Plessey Automation and transferred to Poole, Dorset. In 1968, Telecommunications Research needed a computer to automate Plessey's mainstream communication business, and I became the first hardware engineer.

Unlike batch processing with the IBM 360, computers for telecommunications need five decades of system reliability to match the existing networks. The result was the PP250, the first Capability Limited Computer.

The world was more straightforward then as the dust of WWII still hung over England. Technology moved slowly, and the mechanical analogue products of Alexander Graham Bell and Almond B. Strowger[3] were still in production for Plessey UK at factories in Liverpool and Beeston, where I needed to spend half my time.

Soon, I was flying to meetings at Plessey's pre-electronic, mechanical age factories and visiting Taplow Court, where research

---

3   A.G. Bell and A.B. Strowger patented the telephone and the automatic telephone switch in 1876 and 1891, respectively.

occurred. A strategic corporate investment planned an Industrial Strength Computer to deliver decades of software reliability needed to support a digital telecommunication business. It became the first commercial capability-based computer called PP250. Since disclosure in 1972, the PP250 remains the unique archetype for extreme capability-based computing[4], the epitome of a Church-Turing Machine[5].

For me, PP250 was the starting point for trusted software abstractions creating industrial-strength Computer Science[6]. An Industrial Strength Computer guarantees trusted software. Continuously achieved trust in a dynamic computation is a continuous, repetitive cycle, a scientific process to catch every individual error *'red-handed'* on the spot. The laws of the λ- calculus dynamically bind individual digital abstractions to hardware, using data and function-tight capability limits, uniting hardware and software as two sides of one idea, as expressed by the Church-Turing Thesis[7].

Once a week, the four-seat company plane picked me up at Bournemouth Airport before collecting any Directors, including Michael Clark, son of Plessey's founder, or other Directors on route to distant meetings. Michael Clark ran corporate planning, and the PP250 was his big bet.

Sometimes, I met the Chairman of the Board, Field Marshal Lord Harding. Previously, Lord Harding was the British Governor and Commander-in-Chief of Cyprus during the 1950s when the terrorist guerrilla group, known as 'EOKA,' fought for the union of an undivided Cyprus with Greece. Both were engaging individuals who appreciated the power of innovative technology and the opportunity

---

4   Capability-Based Computer Systems by Henry M. Levy, published by Digital Press in 1984, remains the best comprehensive survey capability-based hardware and software systems. Visit https://homes.cs.washington.edu/~levy/capabook/

5   A Church-Turing Machine implements computer software that obeys the laws of the λ-calculus as defined by the Church-Turing Thesis.

6   Four papers published at the International Conference on Computer Communications (ICCC), October 24-26, 1972, in Washington D. C. including Fault Resistance and Recovery within System 250. ICCC USA 10.1972

7   Defined by the work of Alonzo Church and Alan Turing in the 1930s the thesis postulates that mathematical functions are equally computable by humans, by Turing's α-machine, and by Church's mathematical λ-calculus.

and responsibility presented by inevitable, endless, worldwide digital convergence. Digital convergence is the reconstruction of Civilisation through computer technology, global communications and pervasive software automation.

One morning, Lord Harding explained how helicopters dramatically changed his fight with guerrillas in mountain warfare. On another trip, Michael Clark was intrigued to hear how we spent his five-and-a-half-million-pound investment in the future[8]. Both were fascinated by the digital convergence of society with computer software and digital communications.

Motivated by these meetings, I always strive for reliable computer science as a design for a dependable 'media' platform. Protecting the media in the basement of computer science simplifies and democratises living in Marshall McLuhan's global village[9] because the networked computational platform is functionally safe for the progress of Civilisation. Trusted software, confidential data, and dependable functions are the first three essentials of fault-tolerant, industrially strong computer science. It is a monumental form of cyberspace. It is both a fail-safe and a future-safe computer network governed by citizens in democratic digital societies—a platform to protect the hard-won freedoms of democracy through individuality.

The danger of an Orwellian endgame in a digital society is real. Threat vectors flow from outdated, centralised, binary computers to undetected cybercrime and spying, uncivilised digital corruption, and branded dictatorships. Binary computers only lead to digital enslavement as Artificially Intelligent Malware remotely hacks superuser privileges with another level of Ransomware attacks.

Only when trusted software abstractions offer stable evolutionary software locally will the colonization of cyberspace achieve a steady, reliable form of cyber democracy. Its vision demands Charles Babbage's *Infallible Automation,* which he demonstrated circa 1840—a machine for fail-safe programmed function abstractions.

---

8   Over $75,000,000 in 2019 terms.

9   Herbert Marshall McLuhan, (1911 – 1980) Understanding Media: The Extensions of Man (1964). Coined the terms 'the Global Village' and 'the Medium is the Message.'

Functional digital objects owned by and faithful to individuals rather than digital slaves indentured to centralised monopolies, government dictators, foreign enemies, or common criminals. A Dream Machine detects every functional error by protecting private data, allowing no hacks, detecting malware on contact, flagging deep fakes, preventing outsider attacks, outlawing superuser operating systems, and avoiding centralised dictatorship.

This Dream Machine offers scientific equality and digital justice for all. Inevitably, power emanates from individual citizens, encouraging cyber democracy to emerge worldwide. It is a vital, compelling vision to implement the Dream Machine that motivates my life, and this first book in a trilogy explains the vision, the why, what and how of universal Cyberspace. Book two describes the consequence of a failure of society to act. Book three extolls the advantages of a democratic digital platform, unthreatened by technological shortcomings.

Using monolithic General-Purpose Computer Science, monopoly dictators and cybercriminals have the winning hand. It must change. Global software as trusted function abstractions must necessarily emerge but cannot with outdated binary computers. The necessity is unrealised and seems as distant a dream as it was for Ada Lovelace 200 years ago when she composed and explained the first everlasting function abstraction based on flawless mathematics.

This Dream Machine uses Capability Limited Addressing and Object-Oriented Machine Code, pioneered by PP250 under the eye of Maurice Wilkes. These are the two most important discoveries for Industrial Strength Computer Science. Sadly, the PP250 remains unique. Ignored by mainstream computer suppliers who monopolise the semiconductor industry and limit competition. Object-oriented machine code remains nascent, and with it, the architectural improvements of a Church-Turing Machine for validated, trusted, functionally calibrated industrial strength computer science.

None of this was apparent in the beginning. When Professor Maurice Wilkes, from Cambridge University, became our principal advisor in 1969, he immediately introduced Capability Limited Addressing instead of shared memory address instructions. Thus,

PP250 followed the paper published by Jack Dennis, Earl van Horn,[10] and Bob Fabry's[11] dissertation on the Chicago Magic Number machine. Taplow Court, headquarters for Plessey Telecommunication Research, led the company-wide direction. The daunting requirement for perfect software reliability drove every decision, and we got to work.

Rotary phones chose analogue circuits to connect customers worldwide through various electromechanical relay switches. I vividly recall stepping with care through heavy metal factory workshops to meetings with networking experts in remote office blocks. It was clear I had lots to learn. My mentor, Jack Cotton, led the system design team at Taplow Court, a beautiful English country house converted into a research park, like Bletchley Park, where Alan Turing worked after returning from his seminal time with Alonzo Church at Princeton University. Led by Turing and accelerated by the Bombe[12], the team at Bletchley Park discovered the secrets of the Nazi Enigma code and shortened World War II.

However, his seminal work with Alonzo Church lay fallow. It was ignored after John von Neumann proposed the overstretched binary computer (without the λ-calculus) immediately after WWII. Alan invented the Turing Machine while studying the Entscheidungsproblem[13] with Alonzo Church. In computer science, there exist two complementary forces of logic and physics. The Church-Turing Thesis expresses these opposing forces of computer science, so ask oneself why only the Turing machine exists in a General-Purpose, binary computer science. Understanding and correcting this mistake became a personal goal, but time is running out.

I am indebted to those who contributed to the ideas of Industrial Strength Computer Science. It started with the PP250, an extreme

10   'Programming Semantics for Multiprogrammed Computations' MIT, Jack B. Dennis and Earl C. Van Horn.

11   Bob Fabry, pioneer on Capability Limited Addressing and a computer science professor at the University of California, Berkeley. He obtained DARPA funding to radically improved AT&T Unix and started the Computer Systems Research Group.

12   Bombe - Wikipedia

13   On Computable Numbers, With an Application to The Entscheidungsproblem by A. M. Turing. Received 28 May 1936.—Read 12 November 1936.

Capability-Based Computer, and the first realization of the Church-Turing Thesis, where the sources of abstract logic and physical details are scientifically united. A vital balance exists in every monumental scientific tool from the birth of civilisation with the Abacus, throughout the Industrial Revolution, onto the moon's surface, using Slide Rules and pioneering functional programming using Babbage's two clockwork engines.

I am eternally grateful to all involved. Engineering computers to prevent cybercrime requires a machine that removes human corruption using Babbage's ideals of *Infallible Automation*. Charles Babbage's foresight and Ada Lovelace's vision demonstrate the pinnacle of mechanical computer science, only replicated digitally by Church-Turing machines, the Dream Machine of civilised cyberspace.

Sir Maurice Wilkes (1913 – 2010) guided the technical direction and advised me on microprogramming the capability- based meta-machine that adds industrial strength to computer science. Maurice clearly understood the difference between PP250 and CAP. CAP was a *hybrid* machine for software research at Cambridge University[14]. CAP was the second capability computer project started by Maurice for research students, while the PP250 was the first and, so far, the only example of trusted computer science. On another level, the PP250 demonstrated fail-safe, data, and function-tight, qualified industrial strength computer science.

My research Director at Plessey was Mike J. Lanigan. He explained firsthand the Atlas One-Level Storage system that pioneered virtual memory solutions, started at Manchester University. Mike approved my first publications, building my confidence in the scientific alternative to proprietary computer science. Mike understood that industrial strength computer science required the removal of unscientific proprietary hardware, including virtual memory that requires superuser privileges that allow debilitating malware attacks like Ransomware.

Blindly trusting binary computers, the administrator, the shared virtual machines, the dangerous binary instructions, and statically

---

14   CAP-Research project at Cambridge University on Capability Protection CAP computer - Wikipedia

bound compilations allow undetected malware operations beyond all scientific laws. The undetected errors that result cause international conflicts. Such unfair but programmable powers are central threats to a peaceful cyber society. It results in super-intelligent AI-driven breakouts to escape human control.

At Taplow Court, Charlie Repton drove us to microcode fail- safe, data-and-function-tight machine code instructions framing network-wide Object-Oriented Machine code. Object-oriented machine code is significantly better than binary machine code. Named objects instead of shared memory are the foundation of every digital implementation. Uniquely named tokens define safely packaged and carefully shared functions as distributed assemblies. Each one is an independent item in a private namespace. Unlocking objects for access requires a privately owned digital key, an immutable capability token. Keys reference every admissible digital object of a function abstraction, separately identifying access rights and object types. The PP250 design guarantees trusted software as fault-tolerant computers with decades of fail-safe reliability. It quickly achieved the 'five- nines MTBF'[15] demanded by the telecommunications industry.

The application gurus, Martin Goodyear and Alan Lawrence, explained the networked call set-up transactions as distributed, event-driven threads that reliably connect across a PP250 network of Church-Turing Machines using named tokens as Capability Limited Addresses—a natural model for universal computation.

John 'Jack-o' Cotton,' our Captain, was the leader who guided us all. His constant search for the best, as we worked together over three decades on fail-safe technology, and his willingness to ask innocent but probing questions inspired our success. Like his sailing skills, he could tease progress from the lightest wind and survive the worst weather at sea.

Others made vital contributions, including, but not limited to, Mick O'Halloran, Paul Boucobsa, David Cosserat, Graham Wisdom, Mike Lawson, Geoff Stagg, Roger Boom, and more. Together, the team pioneered a scientifically complete Industrial Strength Computer.

---

15   MTBF The Mean Time Between Failures. Typically measured in decades for industrial strength engineering products.

PP250 was a digitally safe solution for the endless future of computer science. It demonstrates reliable, fail-safe software abstractions as fault-tolerant, networked Church-Turing Machines as the trusted foundation of Industrial Strength Computer Science. A global platform that resists malware attacks, prevents hacking, isolates corruption, controls spying by confirming ownership, and avoids centralised dictators sharing access to any individual's confidential data. For digital democracy in the 21st century, removing the relentless terror of centralised corruption and the unfair privilege of dictatorial administration is vital.

Democracy has capitulated to dictatorial cyberspace. Binary computers are outdated by superhuman AI software. Only return to the Church-Turing Thesis for salvation. Instead of dictatorial centralisation, power must originate from the citizen-users in cybersociety. Industrial Strength Computer Science achieves this goal by reversing the dangerous assumptions of von Neumann[16] and timesharing. Restoring the balanced symmetry of the Church-Turing Thesis corrects an unscientific start in WWII and the root cause of cybercrime. Achieving hardened, trusted software as Industrial Strength Computer Science requires inherited mathematical instincts for software to survive. In this context, Turing's (single-tape) α-machine is just a digital gene for computation executing the logic of a λ-calculus namespace, and a Church-Turing Machine is the software survival machine for computer science. Turing's α-machine, the essential binary computer, executes any program but cannot logically scale. Scaled computations require the dynamic modularity of the λ- calculus. But von Neumann, the originator of General-Purpose computer science, ignored the λ-calculus, a critical part of the Church-Turing Thesis.

The PP250 remains unique but a monumental milestone in the progress of computer science, digital convergence, global communications, and democratic society. It is the prototype of

---

16   John von Neumann (1903-1957), a colleague of Alonzo Church at Princeton University when Alan Turing first described his α-machine. After Turing returned to England and Church moved onto the Manhattan Project, von Neumann proposed a stretched and statically shared computer architecture as an alternative to the universal, dynamic scalability of the λ-calculus embodied in the Church-Turing Thesis.

an extreme Church-Turing Machine. The yet-to-be-built 'Dream Machine'. Only then can commercial computers guarantee trusted software for society to colonise and survive in cyberspace. Digital law, software order and democratic justice are vital for the global colonization of hostile digital cyberspace. As such, the PP250 shares the universal model of computation adopted by the Abacus, the Slide Rule, Babbage's two mechanical engines, Alonzo Church's formula, the λ-calculus, and the natural survival mechanisms of nature and life.

For cyber-society to progress, stable yet evolving generations of programmed software must work reliably for time on end. Like the Babylonian Abacus, the ages of civilisations yet to come will depend on trusted functions of fail-safe computer science. Trusted software that survives is the bridge to the future. Like all other bridges throughout civilisation, the functions must last for centuries. Indeed, programs could last forever, as demonstrated by Ada's Bernoulli abstraction when programmed by scientific symbols, the valid tokens of everlasting science However, microprocessors followed the mainframe mistake and ignored cybersecurity, and their low, off-the-shelf cost tolled the death bell on all competitive alternatives, including the PP250. It remains this way, and the lack of competition destroys the future of cyber society. The end of PP250 became apparent, so in 1977, with Jack Cotton and Alan Lawrence, we moved the idea of object-oriented software and capability-limited binding to ITT Telecommunications Research in Connecticut. The ITT-1240[17] became the next milestone in our search for the safe and reliable digital convergence of computers, society, and international networks like cyberspace. There, George Becker, Manfred Lanangbach-Beltz and ITT-Europe adopted the distributed microprocessor design of the ITT-1240. They trusted me with their destiny at Standard-Electrik-Lorenze when, in 1982 two systems went live in downtown Stuttgart and Heilbronn, Germany, for the DBP[18].

---

17   ITT-1240 A fully distributed microprocessor-based telecommunication switching system installed worldwide since 1982. Notes for a fail-safe, fault tolerant software system that survives under unexpectedly high traffic loads and scales gracefully to any size.

18   The German Post and Telecommunications office.

The early Intel 80xx microprocessors are like updated Turing machines—a programmable α-machine without Capability Limited Addressing for the λ-calculus. Microprocessors refashioned high reliability and scalable functionality by encapsulating device-oriented software function abstractions as hardened object-oriented microprocessors.

While the mainstream computer industry ignored software reliability and only expanded word lengths, increasing the shared memory and the size of statically compiled software, the ITT-1240 used thousands of microprocessors to encapsulate, protect and dynamically bind the object-oriented function abstractions of telecommunications. The amassed industry in Florence, Italy, attending the International Switching Symposium in May 1984, was impressed by the results[19]. If proof is needed to remove centralised, autocratic administrator privileges and centralised operating systems, understanding how the ITT-1240 distributes power to individual computations is a fine example.

At ITT's Advanced Technology Centre, I met Tom Love and Brad Cox, two stimulating software developers and fellow travellers who jump-started object-oriented programming for real-time systems. They created Objective-C[20]. Sharing in the global success of object-oriented programming in general and Objective-C at Apple Incorporated is immensely satisfying. Steve Jobs rescued Apple on the back of Objective-C. Objective-C followed from the PP250 as a practical programming language, an alternative to the Object-oriented machine code, but fundamental to following a standalone[21] isolated form of the Church-Turing Thesis.

---

19   ISS'84 System 12: Integration and Field Experience, 1984, Cohen, K. Hamer-Hodges, G. De Wachter, and H. Weisschuh: Proceedings of the International Switching Symposium, 7 -11 May 1984, Florence, session 32A, paper and Becker, G Chiapparoli, R.S., Schaaf, Robert & Vander Straeten, C. SYSTEM 12: SOFTWARE. 59. 60-67.

20   Objective-C is a general-purpose, object-oriented programming language that adds Small-talk-style messaging to the C programming language. It was the main language used by Apple with macOS and iOS operating systems, and the application programming interfaces for Cocoa and Cocoa Touch prior to the introduction of Swift circa 2014.

21   All Object-Oriented programming languages as prisoners of a compiler and an operating system. Unlike Object-Oriented Machine code these languages cannot be

Thus, object-oriented programs, functional programming, and capability-limited addresses remain limited, trapped by the centralised privileges and binary compilation enforced by general-purpose computers. If shared memory and unfair centralised privileges remain, malware and hacking will persist as increasing systemic threats to cybersociety and the future of civilisation.

Only extreme Church-Turing Machines remove the flaws of General-Purpose Computers. A handful of six fail-safe functional 'Church Instructions' is all it takes[22] to replace privileged, proprietary, branded general-purpose computers.

The six Church Instructions mechanise threaded λ-calculus computation by dynamically navigating immutable capability- limited tokens across networked cyberspace. Hardened, named digital shells isolate the object-oriented, machine-coded or otherwise programmed function abstractions. Its systematic Church-Turing architecture hardens the basements of computer science as a secure yet expanding network called cyberspace.

Uniquely named object shells protect each object-oriented computational item as a digital gene using a λ-calculus meta-machine to bind objects, detect malware, prevent undetected errors and remove hacking by locking malware out of the cockpit. It is a model started by the Abacus, where cellular functions frame dynamic logic as symbolic models using λ-calculus function abstractions that survive forever.

The computational variables of the λ-calculus are abstractions behind each Capability Key assembled as an object-oriented namespace that includes immutable Capability Keys to identify specific items, not by location in shared memory but by names in a namespace. Software functions, abstractions as functional objects, any critical data item, all formatted data, each application thread, and the event-driven computations are composed as individual, application-

---

networked gracefully.

22   The Church-Instructions as pioneered by PP250 are: 1) Swap the Namespace, 2) Save and Change to another Computational Thread, 3) Call on and 4) Return from a Function Abstraction, 5) Unlock Access Rights to us a digital object, and 6) Save a Capability Key as a different data type within a function abstraction.

oriented private Namespace. When assembled and maintained in real-time, an off-line compiler is unnecessary. Each private λ-calculus Namespace remains secure in networked cyberspace, managed automatically and democratically by individuals to meet their needs.

Until the release of Objective-C in 1984, the PP250 object-oriented programming, like λ-calculus, remained ignored. These innovations originated with PP250 in the telecommunications industry. Industrial Strength Computer Science is a real-world problem. It demands immediate solutions to unending digital convergence.

General-purpose computers began as shift-driven batch processors that did not need immediate solutions since any batch of programs that failed could run after correction a day or so later. Industrial Strength Computer Science is different. It must work non-stop, as achieved by PP250. However, cyber threats have only grown since General-Purpose computers ignored the Church-Turing Thesis. Unreliable, statically compiled software has expanded cybercrimes into endless, international cyberwars that threaten the future of civilisation. More than this, the unfair privileges of a General-Purpose Computer are the operating system and blind trust. Centralised, branded operating systems and trained *System Administrators* prevent democratic cyberspace from being ruled by and for the people. The death of democracy, as feared by Abraham Lincoln's words at Gettysburg, is again a spectre. Freedom is not free, and monopolies cannot farm citizens as sheep. Governing America by dictatorial digital oppression crushes the Constitution. It will lead to another Civil War, or international one, to decide if America's dream of a constitutional republic will ever come true.

At ITT, constructive discussions about digital convergence included Brad and Tom. They introduced me to Small Talk, and I explained PP250 and object-oriented machine code. They grasped the advantages of encapsulated software and the value of dynamic binding, so they left ITT to focus on Objective-C.

They started the OOP war years between C++ and Objective-C. C++ won for General-Purpose computers after Steve Jobs acquired Objective-C for his Next Computer. At this time, my colleagues,

Jack Cotton and Neil Olson,[23] used Objective-C to develop the most successful trading desk on Wall Street in the 1990s. The Objective-C language was vital to both developments. Apple's resurgence from bankruptcy, led by Jobs, used Objective- C across all product lines. It was the vanguard of the industrial switch to object-oriented programming. The rapid adoption of C++ and Objective-C reconfirmed the need for Church-Turing Machines as a scientific alternative to General-Purpose Computer Science.

Trapped in a shared memory, designed for batch processing, imperilled by dictatorial superusers, and dominated by central operating systems, software remains exposed to networked digital corruption. As such, results fail because hackers, malware, hidden defects, and users who misuse superuser privileges always succeed. Malevolent attacks like Ransomware succeed because unfair hardware privileges exist by default in every binary computer. Branded computers tilt computations against society. The modern manifestation of the Babbage Conundrum drives computer science to achieve the hallmark of *Infallible Automation*. More is said on this subject later.

The work of Maurice Wilkes, Jack Dennis, Peter Denning, David Hanson, David Tennenhouse and Butler Lampson all left lasting impressions. While we agree on things, the means of computer science remain in disagreement. General-purpose computers are dangerous because they use centralised sharing as outdated, out-of-control, repackaged mainframes incurably infected by malware. Shared memory, unfair privileges, and blind trust in downloaded foreign code are the principal causes. In a network, the virtual machine

---

23   The reason Jack & Neil decided at such an early date to adopt the new technology of Object-Oriented design and programming, and Objective-C in particular, was not only a degree of reuse but equally critical was flexibility. Reliability was an issue, and a trading desk system failure could easily lose billions of dollars. Crashes had to be infrequent, and quickly solved.

The three interesting papers are: https://dl.acm.org/citation.cfm?id=625507, https://www.computer.org/csdl/mags/so/1995/05/s5028.html, and an important paper on software https://dl.acm.org/citation.cfm?id=625328

is a gas chamber poisoned by hackers and malware perverting and corrupting computations on a grand scale. The digital tragedy only grows and ends with the unavoidable break-out of super-intelligent, super-destructive malware and dictatorial digital enslavement.

It all led to this book founded on the Church-Turing Thesis and adopting Alonzo Church's universal model of computation to level the playing field of computer science. The Church-Turing Thesis explains the scientific solution. It is essential for a thriving evolution as a prosperous, growing, democratic civilisation. As the universal model of computation, the calculations are genetic, software versions of a digital Darwinian species.

Cybersecurity is now a national priority. Protecting our lives from malware, digital corruption, government dictators, and false news cannot be learned by AI or programmed in advance. It must be instinctive. It must be automatic and built in naturally. It is a digital shadow, a private extension of everyone. An instance of a software species chained to our DNA as strings of infallible automation. Hardened, object-oriented machine code built to atomically and automatically protect our digital integrity as individuals. Mother Nature's universally fair laws must automatically regulate everyone's security and privacy as a digital shadow implemented as personal function abstractions. Our digital shadows are private functional Namespace designed to protect and defend everyone in a Church-Turing implementation of cyberspace.

For software, *Infallible Automation* in a conflicted world means flawless, fail-safe software functionality. I take the idea of nature's spring from a humanitarian and mathematician, Jacob Bronowski. He authored *'The Ascent of Man.'* I use his poetic wisdom by including his poem on the *'Abacus and the Rose,'* which I quote as the dedication for Chapter 5. In those few lines, he captured everything I believe about the hidden force of nature in computer science.

The only machine individuals can share, coexist, prosper, and thrive without fear or favour is Mother Nature's. Nature's hidden spring is essential to the Abacus and the Rose as the λ-calculus. Its vision guides my steps, and I hope the book does the subject justice to inspire a generation of engineers and scientists to pursue *'Industrial*

Strength Computer Science' with the urgent vigour needed. The continued use of the outdated binary computer concocted in the pre-electronic age using a centralised Cold War security strategy hinders international progress by threatening the survival of individuals, industries, institutions, and nations.

Conflicted nations cannot share hostile cyberspace built on von Neumann's assumption of blind trust in shared memory. The digitally converged global platform for cyber-society must protect and enable hostile countries to live together in the digitally converged electronic village, digitally protected beyond the 21st century. Life without undetected malware and lost cyberwars. Life without unpunished crime or technology-based digital dictatorships.

Instead, digital democracy must thrive for millennia. So, in President Abraham Lincoln's words from his Gettysburg Address[24], a *'government of the people, by the people and for the people shall not perish from the earth.'*

Kenneth J. Hamer-Hodges FIEE

Florida, December 10, 2019, updated April 21, 2024.

---

24   Gettysburg Address - Wikipedia, https://en.wikipedia.org/wiki/Gettysburg_ Address (accessed June 17, 2019).

PREFACE

N ot all will read, understand, agree, or enjoy every chapter in this book. The subject is too dry, but skimming the text skips essential details. Instead, I hope to provoke a debate on the future and the urgent need for Industrial-Strength Computer Science. Everyone is involved; damaging society by accident in the pre-electronic age was never cataclysmic, but critical flaws in the vital technology of the electronic age will destroy 21st- century civilised society.

Yet readers must learn the truth about the mysterious two technologies that underpin the 21st century and govern future generations of humanity. Computer hardware and software redefine the eternal future of civilised progress, but the cults of pre-electronic branded computer science only lead to doom.

There are chapters anyone can read without technical skills. The first and last chapters are for all to understand. The message is clear: we head for national and global cyber-catastrophe, but scientific solutions exist. But is there a will to make the changes needed? The advantages will reap huge rewards. Productivity will rise dramatically as development costs plummet. Cybercrime will end, and everyone will gain. Unskilled individuals will become computer-savvy and take charge of their digital destinies. Ultimately, digital dictatorship will end as democratic cyberspace enhances global society.

There are technical chapters, but I do not write them for a technical review because I do not expect the self-proclaimed experts of branded General-Purpose Computer Science to agree with my diagnosis. They are vested in an unfolding Greek tragedy.

The technical chapters exist to confirm, not to define, the dream of a better alternative. The Dream Machine is a Church- Turing Machines. Society must fight for and demand reliable, usable computer technology, leaving the details to experts using the latest alternatives but redirected by a greater force of Civilisation needed to avoid an Orwellian dictatorial endgame.

The branded commercial interests of the General-Purpose Computer industry live in the past. Simple binary computers are outdated. Software progress is reaching superhuman capabilities while binary computers stay stuck in the past. Malware is unacceptable, and hacking must stop because attacks like Ransomware are intolerable for any advanced industrial society. The computer industry is building digital dictatorships. Constraints protecting the national interest as an evolving cybersociety must exist, and we must learn how to foster cyber democracies instead of a digital dictatorship.

So, I hope legislators find this book instructive. They must turn a new page to save democracy from the commercial dictators who rule cyberspace. Otherwise, results will decline to an unacceptable endgame that approaches too soon, accelerated by the arrival of Artificially Intelligent Malware.

Software unreliability, cyber insecurity, unexpected digital catastrophes, industrial collapse following the tragic loss of intellectual property, critical service outages, and continuous software upgrades create other unsolved problems. The outdated General-Purpose Computer is a rickety binary bridge to cyber society. It is unready for the load and cannot sustain nations as industrial cyber societies. For this, cyber democracy and cyber security are vital. Protecting the virtualised but conflicted world is the technology challenge computer science now faces.

General-purpose computers are the enemy. A General- Purpose Computer is unfair by defending the unelected privileges of the supplier. In cyberspace, they are antithetical to constitutional democracy. It changes the grand experiment of a free society and American justice. Law and order cannot be run by super-intelligent malware in Life 3.0[25] when super-human software roams wild. Thus, cyber democracy only exists if computer users as citizens hold the keys to democratic power over data privacy and personal security in cyberspace.

Within a decade, software automation in every conceivable form will dominate life in the globally interconnected electronic village of the 21st century. But the citizen users are not free. To hide the flaws, digital dictators suppress free expression. The centralised privileged suppliers of the pre-electronic age dictate life in cyberspace through the General-Purpose Computer. Instead of freedom, the villagers are tormented by malware and beset by untrusted software. Trusted software must exist by default for cyber society to survive as a progressing civilisation living in cyberspace.

A Church-Turing Machine responds to this challenge; it is the proven way to guarantee trusted software throughout cyberspace, using an electronic age architecture founded on the logical science of the Church-Turing Thesis.

General-purpose computer science began after WWII and evolved during the Cold War. It grew into Jekyll and Hyde[26], an international weapon of mass destruction disguised as a recreational stimulant. The fuss over Tick Tok exemplifies this dilemma. Through digital convergence, this WMD is already in the hands of greedy criminals and hostile enemies. The added threat of a breakout by super-intelligent malware creates scenarios which are too horrific to contemplate.

The singularity, when machines run the world[27], is dangerously near. Any artificial or accidental malware will be unspeakable and unstoppable; criminals and enemies have already begun to use it. The only solution is industrial-strength computer science, making Church-Turing Machines a national priority.

---

25   Max Tegmark, Life 3.0 Being Human in the Age of Artificial Intelligence (about the danger of living with super smart software)

26   'Strange Case of Dr. Jekyll and Mr. Hyde' by Scottish author Robert Louis Stevenson, published in 1886, concerns a London lawyer who investigates his old friend, Dr Henry Jekyll and the evil Edward Hyde. The novella enters the language as the phrase "Jekyll and Hyde,", a vernacular reference to an unpredictably dual nature: both very good, but sometimes shockingly evil.

27   Singularity is the point at which machine intelligence and human intelligence

. Therefore, this book is provocative as part one of a trilogy on what, why and how to save cyber society. It intends to incite a national debate and a planned reaction to a terminal sickness which condemns the American experiment and disenfranchises every citizen living in cyberspace.

At the risk of repetition, each chapter and each book in this three-part trilogy is self-standing with repetitive footnotes to aid the reader. However, I avoid explaining the Byzantine and opaque software security systems created for General-Purpose Computer Science. They are a given—a nightmare not even experts can solve or explain to the public.

Facts on the ground prove they will continue, simply because at the point of contact when software meets hardware, the machine code is blindly trusted. As a result, malware, hacking, and massive data breaches speak for themselves. The debate only necessitates humanity appreciate the impending disaster. The weakness of General-Purpose Computer Science conceals the unavoidable termination of democratic societies. At the same time, the scientific solution framed in 1936 by Church and Turing and explained in this book solves the problem.

To enforce this point, everyone should know I do not worship the dictatorial Gods of branded General-Purpose Computers. I am an atheist, and like Richard Dawkins[28], I believe in Dawkin's *'gene-*

---

cross. The Singularity Is Near: When Humans Transcend Biology is a 2005 non-fiction book about artificial intelligence and the future by Ray Kurzweil. Kurzweil describes accelerating returns and predicts the exponential increase in technologies like computers, genetics, nanotechnology, robotics and artificial intelligence. Singularity as Kurzweil says build machines infinitely more powerful than all human intelligence combined.

28   Clinton Richard Dawkins, FRS FRSL (born 1941) is an English ethologist, evolutionary biologist, author, and atheist. He is an emeritus fellow of New College, Oxford, and was the University of Oxford's Professor for Public Understanding of Science from 1995 until 2008. Dawkins came to prominence in 1976. His book The Selfish Gene, explained a gene-centered view of evolution and introduced the term meme. Other books introduced the influential concept that the phenotypic effects of a gene are not limited to an organism's body but stretch far into the environment.

centred' view of evolution as the universal model of computation and Jacob Bronowski's hidden spring of life. The natural model of atomic life exists as software, as the function abstractions of λ-calculus. This computer is a Church-Turing Machine.

Turing's single tape α-machine is the nuclear design of an atomic digital gene. Using capability-limited addressing to shelter internal workings, this atomic digital assembly becomes a software survival machine for the cellular organs of one function abstract in hostile cyberspace.

It leads back to *Infallible Automation.* The digital equivalent to the idea pioneered mechanically by Charles Babbage. These digital genes are the clockwork of a Church-Turing computer. They operate together without flawed assumptions about shared physical addresses and unfair superuser privileges and thus leave no room for human-inspired errors. Locked out of the cockpit by atomic-level capability-limited addressing, malware and hackers cannot harm the Namespace.

Each Namespace becomes a *'gene-centred'* digital species. A private, protected, λ-calculus Namespace that evolves through a private directional DNA hierarchy of immutable tokens as private pointers using the equivalent universal model of computation refined by nature that creates life from mathematics by following Alonzo Church's λ-calculus.

A Namespace is a software machine, a cyber suit for the safe colonization of dangerously hostile cyberspace. Namespaces protect each digital shadow in cyberspace as an independent living (dynamic) digital context.

Meanwhile, General-Purpose Computer Science searches, in vain, for the ultimate last patch. So, urgently patched and re- patched software upgrades continue. Patching will never end, and crimes and catastrophes will grow. Cyber society cannot flourish this way. Patching means epic cybersecurity is a mirage or would already exist. Instead, instability, monumental human administration effort, successful Ransomware attacks, and the constant overheads to deliver and install critical upgrades continue at a disturbing pace. Cyber insecurity destabilises society, and as progress grinds to a halt, stagnation, discomfort, and digital dictatorship replace civil democracy.

Experts dismissed this doomsday scenario at SOSP in 1977. Institutions and individuals depend on General-Purpose Computers for jobs, profit, or crime and spying, and they all want things to remain the same. It is inexcusable. Ethics alone should lead to considering the harm inflicted on generations yet to come as Artificially Intelligent Malware breakouts take place. Creating a weapon of mass destruction is a crime against humanity. Unavoidably, civilisation stalls as global wars escalate, crafted attacks increase, and accidental cyber catastrophes rise.

Students of computer science must heed the warning signs and reinvestigate the Church-Turing Thesis, Church-Turing Machines, Capability Limited Addressing, functional programming, function abstractions, the λ-calculus, and Industrial Strength Computer Science. Industrial Strength Computer science is vital to the maturation of the industry. It is a massive undertaking and a life-long pursuit with facets that continue to grow as cybercrimes increase and artificially intelligent malware is used in ways beyond human expectation and understanding or as a reasoned human reaction.

Nations, industries, agencies, individuals, and suppliers struggle in this new age of digital convergence, where super- smart software, weaponised cybercrimes, global cyber wars, and super-intelligent malware roam. Only Church-Turing Machines offer the solution for the survival of cyber society. Fixing the cornerstone of our digital future is the prerequisite to constructive progress as a virtualised democracy where a government of the people, by the people, and for the people is a reality. We, the people, must secure the foundation of freedom, justice, and equality if cyber democracy is to function as intended. It requires a reliable software machine defined by science as taught at school. Only then will future generations trust cyberspace and evolve both democratically and scientifically.

For those interested in more detail on the PP250 through published and unpublished working papers and other references, please use tvod@hamer-hodges.us or visit my blog at www.sipantic.com for up-to-date discussions and specialised help. Finally, everyone should support the Foundation of Life[29] movement to prevent A.I. software breakouts, avoid a tragic endgame, and prevent the loss of individual freedoms won at human cost throughout history. Digital technology must be an individual's loyal and faithful servant, not a dystopian digital master.

---

29   The Future of Life is found here: https://futureoflife.org/ai-principles/

Table of Contents

Ada's Anthem for Computer Science. iii

Dedication. v

About The Author vii

Preface xxiii

Introduction. 1

Chapter 1: On Flawless Computers 5

    The Birth of Computers 10

    The Need for Cybersecurity 12

    The Age of Global Electronics 16

    Capability Limited Computers 20

Chapter 2: The Golden Rule 22

    The Perimeter of Computation 27

    Symbolic Decomposition 30

    The Namespace Chalkboard 33

    From Abstraction to Reality 35

    Computational Threads 43

    The Hidden Springs 46

Chapter 3: The Delusion Problem 50

The Integrity Problem 54

The Deep Fake Problem 57

The Trust Problem 60

The Privilege Problem 64

The Detection Problem 69

The Evolution Problem 72

The Privacy Problem 75

The Breakout Problem 78

Chapter 4: The Integrity of a Machine 82

Symbolic Directives 88

Imperative Commands 90

Symbolic Instructions 92

Object-Oriented Machine Code 96

Clockwork Cybersecurity 101

Chapter 5: The Ishtar Gate 105

True-to-Form 108

Survival 112

Encapsulation 114

Mathematical Machines 116

A Thread of a Computation 121

Dynamic Binding 124

The DNA Hierarchy 128

Typed Machines 131

The Clockwork Meta-Machine 133

Chapter 6: Trusted Software 136

Trusted Software 141

Faithfull Software 147

Too Complex to Test 150

Crime Pays, Dictators Rule 152

The Road to Cyber-Security 154

A Church-Turing Machine 157

Object-Capability Systems 160

Chapter 7: Trusted Computers 165

Nature's Model of Computation 170

Computational Meta-Data 173

Functional Machine Types 181

The Fight for Control 184

Secure Machine Code 188

Fraud and Forgery Resistance 193

Namespace Security 196

Chapter 8: The Clockworks of λ-Calculus 198

Undetected Corruption 201

The Foundation Technology 203

An Evolutionary Model 204

A Functional Class 208

The Surface of Cyberspace 210

Transparent Cybersecurity 212

The λ-Calculus Concepts 214

The Typed Access Rights 217

The Frames of Computation 220

The λ-Calculus Variables 223

Computational Threads 227

Chapter 9: Infallible Automation. 229

The Analytical Engine 233

Alonzo Church. 238

A Universal Model of Computation. 243

Trusted Computers. 245

Faithful Computer Science 248

Software Security 250

Innate Immunity 254

Unfair Privileges 257

Algebraic Computations. 260

Industrial Progress. 262

Chapter 10: The Mathematical DNA 265

The Art of Abstraction 268

Chained Abstractions 271

The DNA of Computers 275

Implementation Hiding. 279

Extended Machine Language 285

Separating Concerns 289

Chapter 11: Ada's Endless Software Machine 291

An Object-Oriented Assembly 293

The Hidden Spring. 296

Inherited Fault-Tolerance. 299

Open Ended Capabilities 301

Evolutionary Stable Software. 304

The Paradigm Shift 307

A Scientific Servant. 309

Ada's Endless Vision 311

Chapter 12: The Cults of Cyber Society 314

Endless Convergence. 318

Flawed Technology 321

Spooks and Spies. 324

Malware D-Day 326

Digital Sarin Gas. 328

Click Jack Attacks 330

Opaque Corruption. 334

Chapter 13: On Digital Enlightenment 336

Atomic Responsibilities 339

Social Responsibilities. 342

Monumental Computer Science 345

The Virtual Chalkboard 347

Infallible Automation. 350

Symbolic Addressing 352

Software Modularity 354

The Church Instructions. 356

The Magic of Variables 361

The Language of Abstraction 364

Chapter 14: Colonizing Cyberspace 366

The Natural Language 371

An Enchanted Formula. 374

The Cyber Suit. 378

Machine Conventions. 382

Data Ownership. 385

Software Reliability. 389

Deterministic Garbage Collection. 392

Policy Controlled Cybersecurity. 396

Chapter 15: The Future of Cyber-Society 399

The Fight for Control. 402

The Point of Singularity. 404

The Wild Frontier 406

The Enemy Within. 409

The Secure Electronic Village 412

The Loss of Control. 414

The Level Playing Field. 416

Trusted Government. 418

Industrial Strength Software. 420

Serving Society 422

The Dream Machine. 423

The Final Words. 425

Table of Figures

Figure 1: Black Box Object-Oriented Machine
Code In A Church-Turing Machine 37

Figure 2: The Hidden Springs In
A Λ-Calculus Abstraction. 40

Figure 3: Scientific Expressions As Symbols
In A Λ-Calculus Namespace 93

Figure 4: A List Of Capability Keys Defines
This Example Namespace. 98

Figure 5: The Bernoulli Expression Evaluated
By Ada Lovelace 103

Figure 6: The Mathematical Dna
Of Decimal Abstractions. 106

Figure 7: Technology Stacks Of General-Purpose
Computers Vs Church- Turing Machines. 157

Figure 8: Comparison Of Cyber Security Models. 161

Figure 9: Pp250 Micro Code Linkage Steps. 167

Figure 10: A Functional Steps In The
Universal Model Of Computation 171

Figure 11: Meta-Data That Bind Capability Key
To Namespace Slots Onto Context Register. 175

Figure 12: The Capability Context Registers
Used By The Pp250 178

Figure 8: Church-Turing Frame Of Computation 190

Figure 13: The Three Cases Of
Frame-Changing Mechanisms. 191

Figure 14: The Key Mechanisms Behind
The Λ-Calculus Rules Of Dynamic Binding. 207

Figure 15: Typed Λ-Calculus Concepts Vs.
Capability-Limited Church Instructions 215

Figure 16: Access Rights In Immutable
Capability Keys (Pp250). 219

Figure 17: The Mathematical Functions
Of The Abacus. 224

Figure 18: A Functional Church Instruction
In A Telecommunications Namespace 226

Figure 19: Interesting Example For
Following The Processes Of The
Analytical Engine Documented By A.a.l. 235

Figure 20 A Private Telecommunications
Namespace Example. 240

Figure 21: A Directional Dna Chain
Of Enteronly (Eo) Capability Keys For An Abacus 273

Figure 22: The Expression Of Bernoulli Numbers (B) . . 281

Figure 23: Note G, The Loop Of Ten Mathematical
Expressions In Ada's Program 283

Figure 24: A Functional Telephone Connection
Request Above With The Assembled
Church Instruction Below. 293

Figure 25: The Machine Code Example. 295

Figure 26: The Instructions Format Changes
Between Computers 354

Figure 25: Pp250 Capability Context Registers 355

Figure 28: The Church Instructions Of The Pp250 360

Figure 29: The Register Conventions Of The Pp250 . . . 383

Figure 30: The Capability Structure
Of Deterministic Garbage Collection 394

Figure 31: Migration To The Electronic
Village Of The 21st Century And Beyond 403

INTRODUCTION

At first, malware was just an ordinary programming bug, a quickly corrected mistake. In those early days, computers and software were designed and built from scratch. Everything was homegrown because nothing was off the shelf. The age of global electronics and software applications was undeveloped. Computer designs and software started with nothing except a trusted slide rule, pencil, and paper.

Programs were all homegrown and initially hand-coded in machine code. Each program statement defined the binary format for raw Turing-Commands in the computer's branded format. Software unreliability was not a concern.

Since then, communications, computers, and software combined through digital convergence have become vital tools, fully merged into modern global society. Yet, Industrial Strength Computer Science remains obscure. The deadly side effects of malware and hacking systemically interfere with General- Purpose Computers. Earlier design flaws grant unfair digital privileges to suppliers, criminals and enemies. Cyber society polluted by undetected malware corrupts cyber society. Unchecked criminals and international enemies usurp the centralised operating system of the outdated binary computer, destroying the fabric of civilisation at every level.

As semiconductors continue to shrink and life-supporting applications expand, digital convergence redefines life as a toxic mixture of super-smart malware, corrupt administrators, digital dictators, and exposed society. Digital computers control life, but

General-Purpose Computer Science has failed to keep pace. It remains the same backwards-compatible, pre-electronic age architecture designed after WWII. It no longer satisfies the national requirements of citizens in Industrialized Democracies.

Semiconductor technology blew the doors off the Cold War era of locked computer rooms where batch processors were isolated from outside interference. The well-staffed locked rooms evaporated as the PCs and smartphones reached out to everyone worldwide. The digitally converged network of the electronic age changed everything, undermining the cornerstones of civil society. The new age offers hope for a better future, but not unless the scourge of undetectable malware and international crime ends.

Malware and hackers attack, infect, and corrupt every law- abiding individual in every civil society. The converged digital platform using outdated General-Purpose Computers is inadequate for today. Using shared memory, unchanged and unsafe since virtual memory began, is plagued by outside interference that only grows. Corruption and conflict originate from unscientific privileges, starting with unguarded shared memory in binary computers.

Data theft, digital spying, fraud, digital corruption, and the continuous impact of cyberwar deconstruct civilisation by destroying democratic society. Subverting law and order within and among nations because cyberspace bypasses the institutions of democracy and good government.

Democracy is fragile, while freedom is not free. Trust in society depends on law and order. When justice is missing, trust evaporates, and society degenerates. Patched software upgrades cannot fix the unfair hardware privileges inherited by General- purpose computers from von Neumann's WWII architecture. Identity-based access control matured for isolated timesharing computers during the Cold War fails when applied to networked cyberspace, filesharing, and instant downloads. For civilisation to survive, trusted software is an unavoidable necessity.

Cyber society now depends on trusting cyberspace. It means calibrated, qualified cyber-security, measured and certified by government standards. It is crucial to the electronic age. The future deserves far more than a patched-up batch processor recast as the latest General-Purpose Computers can offer.

Industrial Strength Computer Science automatically measures software reliability and guarantees trusted software exists by default. Errors are detected, not ignored. *Infallible Automation* is a prerequisite for the life-supporting software of the digitally converged electronic village.

Trusted software demands that each machine command be validated and verified as fail-safe at every step. Science, once evaluated, lasts forever. Likewise, the Abacus and the Slide Rule are eternal because mathematics is eternal. For civilisation to thrive, the software machines of the 21st century must last flawlessly for generations on end.

CHAPTER 1

On Flawless Computers

*'The Analytical Engine has no
pretensions whatever to originate
anything. It can do whatever we know
how to order it to perform'*

*Augusta Ada King-Noel, Countess of
Lovelace, the Enchantress of Numbers,
friend, and associate of Charles
Babbage and author of the first
documented software program.*

In 1822, Charles Babbage[30] invented the flawless clockwork engine he called the Difference Engine[31]. He engineered this computer to remove all sources of human error from his printed results. *'Infallible Automation'* was his hallmark. The Difference Engine demonstrated perfection by printing results as tables without needing a human typesetter. In 1837, he described his Analytical Engine[32] and went

---

30   Charles Babbage (1791-1871) KH FRS was an English polymath. A mathematician, philosopher, inventor and mechanical engineer, Babbage originated the concept of *Infallible Automation* by removing all human errors from his printed results. His mechanical implementation of the Difference Engine proved the mathematically functional clockworks and his subsequent design for an Analytical Engine was the first programmable computer.

31   Difference Engine see the *'Edinburgh Review'* of July 1834, for a very able account of the Difference Engine.

32   Analytical Engine see Sketch of the Analytical Engine Invented by Charles Babbage,

further. He engineered a programmable computer, a flawless clockwork machine for programmable mathematics. *Infallible Automation* removes all human threats and unfair privileges, including those that disrupt the calculations of a General-Purpose Computer.

He was personally concerned with every engineering detail. He demanded mechanical and mathematical perfection to fourteen decimal places. Had he lived today, Babbage would object to the General-Purpose computer's unfair hardware privileges that corrupt General-Purpose Computer Science. Humans use unfair privileges to hack into machines. They steal private data, perform a wide range of undetected cybercrimes using crafted malware, and leave a dangerous array of undetected programming errors.

Babbage used *Infallible Automation* to remove human involvement through unassailable clockwork functions. His pure, flawless mechanics mimicked the ones we all learn at school. It is the same scientific foundation needed to achieve trusted software. Mathematical science defines the flawless functionality required for endless cyber-dependent civilisations.

Babbage extended his machine to print flawless tables correct to fourteen decimal points. Babbage removed all sources of human error to guarantee his trusted results. His two clockwork computers would faithfully serve society by fully serving mathematics in depth and detail. In WWII, von Neumann had no vision; he just wanted fame by being first. Every mistake made by General-Purpose Computer Science is artificial, instigated by von Neumann's ambition to be first. His expedient shortcut ignored the λ-calculus. The resulting exposed machinery encourages human interference, errors, accidents, and deliberate sabotage.

Timesharing amplified the first mistake by adding unfair administrator privileges for the centralised operating system services and proprietary concoctions for interworking. Suppliers extended the master-slave arrangement to hide virtual memory hardware

---

by L. F. Menabrea, of Turin, Officer of the Military Engineers, from the Bibliothèque Universelle de Genève, October 1842, No. 82, with notes upon the Memoir by the Translator, Ada Augusta, Countess of Lovelace, the visionary mathematical daughter of the poet Lord Byron.

and became the branded foundation of all binary General-purpose computers. Each branded cult is unscientific. They are organised by the monopoly supplier using these unfair administrator privileges to suppress malware and users in a digital dictatorship.

Consequently, General-Purpose Computers have tilted playing fields favouring the dictator, undetected crime and criminal henchmen. The citizen is disadvantaged because cyberspace is unjust. Errors remain undetected. All the flaws emanate from the von Neumann architecture, an assumption of blind trust in monolithic software and the opaque security of Cold War timesharing founded on Identity-Based Access Control[33].

Not long after Alan Turing left Princeton University, von Neumann ignored the work of his Princeton rival, Alonzo Church and stretched Turing's α-machine into the von Neumann architecture soon adopted by General-Purpose Computer Science. His shared memory architecture ignores the λ-calculus, Alonzo Church's elastic half of the Church-Turing Thesis—the blind assumption of perfect software allowed von Neuman to stretch Turing's α-machine and share the binary address space between all machine instructions. Thus, von Neumann's architecture scales monolithically by destroying scientific integrity. Consequently, a compilation step is required to share accessible memory as a virtual machine.

For the mainframe age, before the personal computer, it worked, but networking changed the rules. Network downloads compromise the static assumptions of the virtual machine, and unavoidably, under the dynamic stress of timesharing, shared physical addressing fails. Sophisticated malware used in Ransomware attacks exploits this design flaw to seize control of centralised operations by usurping the operating system.

The von Neumann architecture only scales by increasing memory word length and memory size. It led to page-based virtual memory, proprietary paging hardware, superuser privileges, central

---

33   Identity Based Access Control also known as Identity-based security limits access to computers, products and services by authenticating individuals. Centralized monitors set up by authoritarian administrators check system calls to the operating system to constrain access rights using a matrix of credentials matching users to resources. Software downloaded from the internet too easily bypasses these selective checks.

operating systems, and centralised digital dictatorship. Consequently, Babbage's *Infallible Automation* is impossible to achieve. The baggage of timesharing only worsens von Neumann's dangerously flawed assumptions. The effort to avoid these flaws only adds administration overheads and more dictatorial constraints on programmers, languages, browsing, and communications.

When von Neumann first wrote the code, he could rewrite it in his head, but soon, that was impossible. Still, business software was home-grown throughout the Cold War and quickly patched up by expert on-site staff. The staff recompiled a brand-new build for the next shift. At the same time, Identity-Based Access Control allowed Information Technology (IT) experts to protect their work privileges from departmental clients.

In the PC age, the von Neumann strategy went downhill. Programs with undiscovered errors increased, leading to unexpected crashes by browsing strange servers, downloading unknown code, and importing unknown data on the fly from the far side of the world. The risks are enormous and fraught with danger. Malware now imported on the fly was beyond von Neumann's goal to be first, but not beyond Alonzo Church's vision. It remains inevitable that undetected malware silently infects every interconnected General-Purpose Computer. The blindly trusted virtual machines of statically compiled software are easy to hack. The authoritarian super user needed to administer every computer is easily confused. Ransomware will always succeed, given time and a determined enemy.

The correct assumption for the electronic age and the endless future is that *'software-cannot-be-trusted.'* Not just because the medium is unreliable and inconsistent, nor because it is complex, and not because it is impossible to evaluate thoroughly. All of this is true. However, the deep and dangerous reason software cannot be trusted is that, in Ada Lovelace's words, the *'weak [human] mind'* is behind every program. As Babbage understood and von Neumann ignored, purging computer science of human frailty is vital for the *Infallible Automation* of mathematics as computer science. *'Infallible automation'* is not an option using von Neumann's shared memory architecture. However, it is essential for the trusted integration of global cyberspace, which is critical for civil society in the 21st century and beyond.

Every machine command is the product of human inspiration, curiosity, direct and indirect competition, pure greed, and naked envy. Thus, every machine command is a threat. These same human shortcomings of students and colleagues inspired Babbage in the summer of 1821. Poring over hand-calculated, error-prone astronomical tables calculated by students and colleagues, he exclaimed to his friend Sir John Herschel, *'I wish to God these calculations had been executed by steam.'*[34] Then and now, the *'unerring certainty of machinery'*, in Babbage's words, must be harnessed to solve humanity's accidental and criminal weakness.

John von Neumann diverged from Charles Babbage's *'Infallible Automation'* and the trusted mathematical clockworks of λ-calculus created by Alonzo Church. By stretching the Turing machine, computer science lost the means to scale atomically as a scientific future-safe clockwork solution. The architecture sentenced the General-Purpose Computer to locked rooms, with resolute IT support staff and homegrown software of the Cold War era.

For the digitally converged network called cyberspace, it is unacceptable, impractical, and inefficient. But the cybercrime result is intolerable. Networking exposes the software to undetected criminal corruption, opportunistic crimes, and unexpected breakdowns.

Unfair freedom exists to secretly spy, hack, steal, and attack the hard work and intellectual property of others. In the hostile, conflicted world of the 21st century, software is unsafe because humanity is at war, and enemies are at work. Corruption in the 21st century is rich with ever-expanding uncivilised opportunities created by a pre-electronic age, stand-alone architecture in a digitally converged world.

---

34   Oxford Dnb Article: Babbage - www-history.mcs.st-and.ac.uk, http://www- history.mcs.st-and.ac.uk/DNB/Babbage.html (accessed June 18, 2019).

THE BIRTH OF COMPUTERS

Babbage made no flawed assumptions. Indeed, trusted computers started far earlier with the birth of civilisation. Ever since the Abacus, society has improved. It all began in the walled and gated, well-defended City of Babylon. Here, driven by trade along the Silk Road, the Abacus evolved as a mechanical computer for simple arithmetic. The Abacus was a natural selection over piles of stones. Time and again, the Abacus served markets worldwide perfectly for generation after generation. It is an everlasting monument to the dawn of civilisation and the birth of flawless mathematical clockwork.

Later, circa 1625, the equally flawless Slide Rule fuelled new feats of mechanical mathematics so powerful it enabled the Industrial Age and beyond. The perfect Slide Rule engineered the steam engine. Then, the Wright brothers learned to fly, and within a lifetime, NASA, through Buzz Aldrin, landed *'Eagle'* on the Moon[35]. The accuracy and reliability of the infallible Slide Rule enabled the ultimate achievement onboard Apollo 11 and Neil Armstrong's small step as a *'giant leap of mankind.'*

Civilisation develops through knowledge and understanding, driven forward by science and technology. However, progress depends on reliable, dependable solutions. Poorly engineered solutions lead to future disasters. Engineering is the art of the possible. Engineers

---

35   Buzz Aldrin took a six-inch pocket slide rule with 22 scales, a Pickett Model N600-ES to the Moon, essential on that first landing to check the overloaded computer functions. The N600-ES sold for $10.95 in 1969.

must safely engineer solutions for the future of an enlightened society. Tragically, General-Purpose Computers fail to meet this hallmark. Solving the Babbage Conundrum and achieving Infallible Automation is essential for civilisation to progress in the Information Age.

Malware and hackers interfere by destroying the reliability and the faithful service of pure mathematics taught at school. The General-Purpose Computer is capricious. Unlike the Abacus and the Slide Rule, Dr Jekyll and the killer Edward Hyde co-exist, hidden within the binary computer. The split personality lacks the ability of *Infallible Automation*, as found in the clockworks of the Abacus and the Slide Rule. Infallible automation denotes industrial strength. The clockwork rules of science flawlessly survive the tests of time and nature's uncertainties. It is vital. Every future generation depends on this.

Stolen secrets, disrupted lives, destroyed futures, disabled industries, misled governments, and abandoned democracy are bad enough, but these undetected crimes allow criminals to go free, and nothing improves. Corrupt software and dictatorial administrators make equality and justice unattainable. The virtualization of industrial society using General-Purpose Computer Science only leads to corruption and branded cyber dictatorship.

When von Neumann overstretched Turing's α-machine, he ignored the λ-calculus. He cracked and broke the science defined by the two balanced sides of the Church-Turing Thesis. Babbage would be disgusted. John von Neumann riddled his architecture with unscientific privileges, facilitating crafted crimes, simple mistakes, and deliberate attacks to go undetected. The General- Purpose Computer changed the world with dangerous, unregulated side effects that always hurt civil society. These human-inspired weaknesses are the same ones Babbage resolved a century earlier. Such misunderstandings led to every war throughout history.

THE NEED FOR CYBERSECURITY

J ohn von Neumann's ambition hastened the adoption of his computer architecture[36] through a lack of patent cover. The urgency of ending WWII that became the Cold War and the Moon Race all too quickly led to the centralised, time-shared General-Purpose Computers, locked in cloistered rooms staffed by cult experts skilled in branded versions of von Neumann's overstretched computer architecture. By 1965, the central operating system needed to manage virtual memory and timesharing of the binary computer. It led to centralised proprietary designs, various rings of privilege, overly complex operating systems, and the almighty superuser.

None of this baggage is necessary or scientifically ordained. None of this exists in the Church-Turing Thesis, in Babbage's flawless engines, or in nature's mathematical clockwork of life. Any proprietary baggage tilts the playing field to serve monopoly suppliers, criminal interests, or others engaged in spying. The lack of security exposes the software to undetected attacks, easy theft, and crafted enemy sabotage.

---

36   Working at the University of Pennsylvania on the EDVAC project, von Neumann wrote an incomplete First Draft of a Report on the EDVAC. The paper was prematurely distributed preventing patent claims by the EDVAC designers J. Presper Eckert and John Mauchly. In von Neumann's architecture data and program share the computer's memory in a common address space. It architecture is the basis of the General-Purpose Computer also called Stored Program Control or SPC. It single-memory, stored program control architecture of von Neumann was in fact the hard work of Eckert and Mauchly.

Cybersecurity is not an option. Cyberspace must preserve the individual, defend privacy, prevent spying, detect theft, and uncover corruption, or civilised industrial society will undoubtedly fail. It is essential to acknowledge this problem and do so before the point of singularity and Life 3.0.

Babbage's *Infallible Automation* is the only way to sustain international law and order in the age of digital convergence using globally interconnected electronics and virtualised services. Von Neumann's assumption of blind trust in software perfection was always an unacceptable, unreasonable, and unsustainable shortcut. It is for these reasons Babbage would complain. Removing the source of human error was his *reason d'être* for the clockwork Difference Engine and his flawless Analytical Engine. Axiomatically, computer science should implement mathematical science faithfully. The public expects this. They understand undetected cybercrime. Admitting to the flaws in General-Purpose Computer Science is the essential first step for the progress of cyber civilisations.

Anything less hides impending disasters caused by engineering, business, and social tragedies. Best practices and monolithic software cannot scale beyond isolated virtual machines in locked rooms. General-purpose computers need virtual memory to grow, but more memory only increases the risk of undetected corruption and fails to deal with flawless communications. Its path is downhill. It cannot satisfy the precise, ultimate demands of virtualised competitive cyber societies evolving independently as 'un' United Nations.

For the 21st century and beyond, we need a long-term solution. The risks of General-Purpose Computer Science, machines designed as batch processors locked in rooms staffed by trusted experts with security-cleared passes, cannot be accepted. These rooms were without network connections; deep in the NSA, they remain so.

Plagued by inevitable, incurable, unavoidable human conflict, any digital download that depends on blind trust is not just irrational. Instead, it is stupid. Global cyberwar only grows while law, order, and speedy justice decay. As increasingly clever software opens ever better criminal opportunities, the unfair results suppress society. The infrastructure of the epoch is letting civilisation down.

Instead, this infrastructure must be an everlasting monument to civilisation's creativity. It must be like the Abacus, the Slide Rule, and Ada Lovelace's Bernoulli function she wrote 200 years ago, faithfully surviving generation after generation. If not, democratic society will progressively disintegrate into old- fashioned corruption, returning society to the Middle Ages. General-purpose computers cannot defend themselves, so they certainly cannot defend civilisation for decades, becoming centuries then millennia. Instead, corrosive attacks, global cyber wars, and branded dictatorships gain control. Constitutional democracy evaporates as digital crimes turn into corrupt, centralised tyrannies that traumatise life and destroy society.

The problem is that blind trust leaves malware and cybercrimes undetected and digital corrosion hidden. Disabled by cybercrimes, individuals, industries, and societies fail to perform. All because the attacks are undetected and ignored at the point of contact, at the place and time when easily detected, red-handed on the surface of computer science by the λ-calculus. For too long, the root cause remains unaddressed, and von Neumann's assumption of blind trust is unchallenged by mainstream computer scientists.

Outsiders see the folly; computer science lost its way in a race that led to Hell. General-Purpose Computer Science is unfair to individuals, society, and the cause of democracy. Easily acquired malware infiltrates private systems, using unfair powers to spy, steal, and hack a massive data breach. Equal justice for all does not exist when unidentified strangers and enemies take over control. Corruption in cyberspace is antithetical to the progress of advanced civilisations; this cannot satisfy industrial society in the 21st century.

Enemies, gangsters, and criminals roam free in cyberspace using unearned privileges that stunt and abuse the civilised progress of society. Malware and hackers exploit unfair opportunities by navigating General-Purpose Computer Science's exposed, interconnected weaknesses to achieve a personal, financial, or competitive advantage. Another generation of malware bypasses the byzantine network barriers thrown up at huge software costs and performance overheads. Blind trust misfires where it hurts most on computer science's computational surface, strictly where scientific security matters most and offers the maximum advantage.

Patched software upgrades are urgently released for installation (or not) every month. Despite this, new gaps, cracks, and voids emerge on the surface of computer science. This digital frontier is like the untamed Wild West. The digital frontier is where hardware meets software and good interacts with human evil. It is the scientific dead centre of the Church-Turing Thesis.

By ignoring the λ-calculus, new attack vectors appear after every upgrade because guarded modularity is non-existent. The absence of protection makes the details of complex software impossible to evaluate and regulate. The ever-changing, dynamically interacting functions in a monolithically composed compilation interact unreliably through blind trust. The pre- electronic age assumption ignores error detection, and with open sharing, the General-Purpose Computer guarantees that cybersecurity remains forever unsolved.

THE AGE OF GLOBAL ELECTRONICS

The von Neumann architecture is from a different age, a pre-electronic period when less than a handful of computers were all built from vacuum tubes and delay lines, not transistors and random-access memory. When computers lacked enough 'random-access memory' and unit costs were so high, memory sharing was essential. In the pre-electronic age, experts outnumbered computers, but now computers dangerously outnumber innocent citizens. The table turned. Estimates say over 1,000 computers automate each fighting member of the armed forces; I have four in front of me as I type, all far more potent than when von Neumann's shared memory architecture was deemed essential at the end of WWII.

General-Purpose Computer Science emerged from a world when hardware and software were homegrown and global networking did not exist. Downloading software from anywhere and everywhere exposes computers to sudden outside attacks. John von Neumann's architecture fails to detect software errors and encourages undetected malware attacks. Hackers, criminals, and enemies silently steal and cheat, and passive General- Purpose Computers take no notice.

Since the Cold War, semiconductors have turned the world upside down. Society now depends on software to run lives, industries, communities, and the nation, but as a foundation for 'virtualised democracy', critical essentials of equality, law, order and justice are missing. Software unreliability and superuser abuse are deadly, and gradually, as software automation expands, digital dictators take over. The impending catastrophe is an incompatibility between the needs of civil society and the design of General-Purpose computers to do that job. The tarnished success of monolithic software inhibits rapid progress; the primitive, pre-electronic age security system is just software called Identity-Based Access Control. However, software alone cannot logically check a download, leaving only downsides for stability in the dynamically interconnected, international future.

Technology now collides with the needs of society. Malware must be locked out of the cockpit in advance to prevent terrorists from invading any computation. It is the same problem faced by the airline industry after 9/11. The cockpit can only be made secure by door locks and keys. In this upside-down world, von Neumann's architecture is outdated, and despite all the industrial effort, General-Purpose Computers remain backward compatible. It is stuck in the past, married to an overstretched solution, and using the Cold War software security strategy of Identity-Based Access Control. It ignores the demands for lock- and-key security, trusted software, and inherently secure networked computers.

Babbage's hallmark of *Infallible Automation* must exist on the surface of computer science as software touches hardware at the dead centre of the Church-Turing Thesis using locks and keys to access or enter every digital object. It is scientific engineering easily understood by individuals, industries, and democratic nations to survive into the endless future.

The Information Age of global cyberspace needs Babbage's Industrial Age vision and Alonzo Church's λ-calculus science— Turing's α-machine nor von Neumann's architecture is sufficient. Instead, they are distractions away from the Church-Turing Thesis. Cyber survival depends on the Church-Turing Thesis as the perfect, fail-safe network infrastructure where software is impervious to corruption. A reliable and impartial scientific platform devoid of superuser concoctions and supports all citizens with equal justice for all. It is a level playing field for cyber civilisation to excel. A place where experts, citizens, criminals, and enemies extended by Cyberspace remain digitally separated and held private and secure. A place to safely survive in a hostile, human-inspired environment called cyberspace.

Equality must exist in cyberspace for software to safely support a cyber-dependent Democracy. The programs must be pure, devoid of outside interference, careless and corrupt human activities, enemies and terrorists. The citizens, not the suppliers, must transparently control data privacy and cybersecurity programmatically by owning confidential keys to their information. A level playing field must survive flawlessly for generations to serve society as a bridge to the future without corruption, shortfalls or bias. Trusted computer science, proven by the clockworks of Charles Babbage, reconfirmed digitally by the PP250, originated by the Babylonian Abacus, then the embedded algorithms in William Oughtred's Slide Rule, and ultimately by mathematical statements that last forever as programmed by Ada Lovelace's Bernoulli abstraction.

Using the branded architectures of von Neumann's binary computers allows humanity's dark side to triumph using undetected malware and crafted outside attacks from the far side of the world. Unguarded access rights to shared memory enable unidentified strangers to put their digital thumbs on the scales of equality and justice to upset democratic society. Computers are so dangerous that ransomware freezes targeted sections of society overnight, as with the UK National Health Service and organizations and cities in the USA. Cyber-powered life stops dead as computers are digitally frozen, robbing industries worldwide of private intellectual property. Equally, clickjacking attacks rob innocent citizen-users of peace, stability, and success.

The downsides are overwhelming; it started with PCs and networking decades ago. Millions of citizens in growing numbers are all compromised by massive data breaches that cost the service supplier over $150 per identity and over $4 million per attack. These numbers are from IBM's 2018 Cost of Data Breach Study and exclude specialised attacks on high-value targets. The study concludes that the faster any attack is detected, the lower the costs. It is for this reason Capability Limited Addressing is so effective.

The hordes of stolen identities and security credentials jeopardise access to critical computerised services for months, if not years. IBM reports that under the best conditions, it takes 7- 8 months for detection and 2-3 months for resolution. However, the resolution is inadequate given the circumstances. The painful recovery process leaves individuals enslaved through lost credentials, captured by business monopolies operating as digital dictators. Life is unfair this way. Cyber society must extend democracy, not replace it with dictators.

It is impossible without digital equality, transparent law, order, and speedy justice. Equality begins on the surface of computer science. Capability-limited addressing reduced the delay to *'on-contact'* with any error and *'immediate recovery'* without data loss or corruption.

Even better, international diplomacy improves through digital transparency. Cyberweapons are detected instead of ignored. Affairs cannot degenerate into pretentious warnings and pompous threats. These threats only lead back to a traditional war, using, in the worst case, the bombs that ended WWII. The pre-electronic age Genera-purpose computer is ruining the future of the 21st century and beyond. It is only a matter of time before attacks succeed faster than patched upgrades can cope. Everything will end up just spinning wheels.

Every citizen in every nation is affected by the flawed pre-electronic age assumption of blind trust in software. The rules of cyberspace must change by updating computers and building on the Church-Turing Thesis. Industrial Strength Computer Science leaves behind the locked rooms of a Cold War mainframe by abstracting lock-an-key security as the object-oriented machine code of a Church-Turing machine.

Multi-programmed global networking was over the horizon when batch processing enjoyed unprecedented growth. However, as soon as networking began, insecurity surfaced, malware and hacking became unstoppable and General-purpose computers needed replacing.

CAPABILITY LIMITED COMPUTERS

Long before it mattered to the mainstream, software unreliability forced the communications industry to explore Capability Limited Addressing. By 1972, the PP250 pioneered Capability-limited computer science as a trusted solution with decades of software and hardware reliability. Furthermore, the ultimate capability-based computer unfolds the supercharged power of the λ-calculus included with the Church-Turing Thesis. Software is protected mathematically by a Church-Turing Machine, encapsulated by a λ-calculus meta-machine, scientific symbol by scientific symbol.

As Ada Lovelace first recognised, these symbols are anything physical or abstract, not limited to numerals as Babbage intended. It is because a Church-Turing machine obeys nature's universal model of computation. Its computational method encapsulates Turing's 'single-tape' α-machine, one function at a time, using Alonzo Church's 'λ-calculus' as an independent application namespace. Think of the application as a species of software with a distinct DNA, defined by a hierarchy of immutable tokens called capability pointers.

It is the universal model of functional computation Alonzo Church called the λ-calculus. The λ-calculus binds software scientifically as individual symbols to Turing's α-machine, where each symbol is scoped and protected by Capability Limited Addressing, protecting individually typed digital objects. Furthermore, and most importantly for democracy, the Capability Keys are tangible, immutable digital assets placed in the hands of private citizens as data owners to extend their individuality functionally, physically and securely into networked cyberspace. At the same time, service providers supply 'λ-calculus function abstractions'. Everything is capability-limited, only accessed through locks and keys.

The λ-calculus functions are object-oriented programs written in machine code as secure digital classes used by private user namespaces. Thus, the nature of computer science changes from a centralised authoritarian master using opaque privileges to transparent, faithful, and loyal servants distributed locally and over the global network to serve individual tasks. It serves the communications-driven, digitally converged cyberspace of the future where data is private and the government of, by and for the people, can grow and prosper democratically.

Using a Church-Turing Machine, every function is infallibly automated. Materializing Babbage's ideal in the global computer network creates a flawless set of λ-calculus machines as Ada Lovelace envisioned, now called cyberspace.

If the flaws remain, malware breakouts will continue. However, worse will come if software improves while hardware stagnates. The smarts of Artificially Intelligent Malware (AIM) make everything worse. The consequence of super-intelligent software breakout as Life 3.0 is unacceptable. Its existential threat to civilisation is now in sight. It sets the point of singularity as a deadline for this generation to replace centralised, dictatorial, General-Purpose Computer Science with the balanced, scientific, distributed crosschecks of a Church-Turing Machine.

Babbage and Lovelace foresaw this industrial imperative 200 years ago. The scientific progress of civilisation depends on the *Infallible Automation* of computer science. The ideals of Charles Babbage and Ada Lovelace define the indispensable ideals for the future. Computers must be flawless, faithful servants of society, not unruly, opaque, capricious dictators of cyberspace.

CHAPTER 2

The Golden Rule

*The saddest aspect of life right now is
that science gathers knowledge faster
than society gathers wisdom.*

*Isaac Asimov (1920 –1992) was an
American writer and professor at
Boston University, known for his works
of science fiction, popular science and
The Three Laws of Robotics.*

When computer science began in Babylon, natural selection picked the Abacus as the pocket calculator of choice precisely because it is mechanically exact and mathematically faithful. It is a dependable clockwork servant of society. Until the age of electronics, citizens trust the Abacus above all alternatives. Babylon, no doubt, evaluated and rejected a general-purpose pile of stones and the intervening millennia proved them right, not once but repeatedly. No obscure skills are needed, and the abacus serves both mathematics and the citizens of society equally and fairly. Thus, the Abacus spread like wildfire, taking trade and civilisation globally.

The Abacus is eternally faithful as a clockwork mathematical framework. It is accepted globally. Beyond doubt, it is an incredible, monumental achievement of the ancient world and a great gift from the ancient past, paid forward to the endless future. As a clockwork machine, it is forever perfect. It is a servant of two masters at once: mathematics and civilisation. The Slide-Rule and Babbage's two mechanical engines, the Difference

Engine and the Analytical Engine, are cut from the same cloth.

Every perfect machine serves a human purpose and society faithfully and in every respect, consistently passing every test of time through *Infallible Automation*. These machines can last forever because their form is not general-purpose but engineered for specific flawless functions. Thus, each software function has a faithful, purposeful, object-oriented digital design.

As such, like pure mathematics written on a university chalkboard, a Church-Turing Machine is the student of universally true mathematical science. For example, Ada Lovelace wrote her function abstraction of Bernoulli Numbers using the pure mathematical expressions of the Analytical Engine. Such scientific functions never age or decay over time. The expressions Ada used survive forever, and the corresponding clockworks of her function abstraction pass all tests of time. Ada's programming language was mathematics. See Figure 23 for a snippet of her machine code with one school room statement for each machine code command.

The software of this well-engineered technology is symbolic. Likewise, Capability Limited Addressing uses immutable digital keys as symbols, such as a, b, c, +, and -, found in Ada's code. Symbols survive reliably and forever when expressed and implemented by scientific methods. It is vital to dependent, trusting society as open, democratic nations on-route to life- supported, infallible automation of every conceivable service. Software implemented as pure mathematics is perfectly understandable, easily explained, and endlessly dependable. It is timelessly secure from ageing, misunderstanding, and outside attacks. It offers equal functionality perpetually for all. Functional software is composed this way; once tested and debugged, it never needs to change! One only needs to read Ada's Bernoulli program, written in Babbage's machine code (but not yet debugged), for proof.

However, software reliability is vital for sustained correctness over time. Critical software must remain, like a bridge, endlessly faithful to serve society beyond the 21st century. Ada Lovelace wrote her program (see Figure 23) in the first half of the 19th Century, but (if correct) the machine code statements survive forever because they are programmed purely by the language of mathematics. The same mathematics learned at school. Furthermore, implementing these same abstractions as data- and function-tight objects leaves no room for malware or hackers. They have no free reign because unfair privileges do not exist in capability-limited computers. It is how two plus two always makes four, worldwide and forever.

Critically wrapping the functions of binary computation as symbols in Capability Limited Addressing enforces Alonzo Church's universal model of computation, creating a Church- Turing Machine for the 21st century and beyond. Unlike the pre- electronic age General-purpose computer, a Church-Turing machine belongs to individuals holding keys instead of dictators with central controls. The formula is a shared constant, while the substituted variables remain private property. General-Purpose Computer Science's centralised, branded privileges subvert the natural rules of science and nature to favour service providers over citizens.

Individuals who use the universal model of computation can confidentially own their data in a Church-Turing Machine. A secure and private λ-calculus namespace belongs to each citizen.

One at a time, individual computations share computers throughout cyberspace as a coherent Namespace, protected as a private digital shadow. Everyone's shadow exists as a lock-and- key namespace secretly, as a cyber citizen, with their particular DNA as individuals clad safely for digital cyberspace. Marshall McLuhan's Extensions of Man is made true for computer science throughout the information age.

Tolerating undetected digital corruption or functional breakdown is a fatal mistake for electronic media in the Information Age. Every virtualised action is a matter of human concern. Therefore, every action must be scientifically pure, fail- safe, and guaranteed as data- and function-tight. When boundaries merge, this becomes impossible, as with the monolithic, statically compiled images used by General-purpose computers.

Software as flawless, faithful, and eternal as Ada Lovelace composed requires Babbage's flawless clockwork. At the same time, these dynamic computations must be data-tight extensions of individual, function-tight abstractions from service providers. Such digital computations are trusted because every digital abstraction is completely media-tight, functionally reflecting each type. A unique symbol as a capability key can unlock specific access rights to clear digital frames as a pure atomic functional expression. It democratises machine code using the notation encountered at school or on a university chalkboard. The scientific proofs learned at school as mechanised by the Abacus, the Slide Rule and a Church-Turing Machine are understood universally, democratizing the mechanics of computer science for all to enjoy.

Trusted software does not exist if corruption breaks out. Malware and hacking are the apparent signs of scientific failures in General-Purpose Computer Science. The Babylonian golden rule for the Abacus is 'atomic-form matches atomic-function.' When form-matches-function computations carry no corruptible baggage, there is also no overhead. The resulting simplicity is like the clockwork Abacus, a mechanical Slide-Rule, Babbage's two Engines and Alonzo Church's λ-calculus.

Consequently, a Church-Turing Machine is a lightweight and high-performance mathematical machine that computes one functional expression at a time, for example, $(a + b = c)$. The Object-Oriented Machine code forms the private *'data-tight'* calculation using a *'media-tight'* function. The calculation depends on the perspective from the opposite sides of the Church-Turing Thesis. Capability Limited Addressing balances the physical to the logical relationship, uniting two views, one of form and the other of function, converting $(a + b = c)$ into a single fail- safe expression executed in isolation like a single tape on Turing's α-machine.

Thus, Capability Limited Addressing straddles the frontier of computer science at the dead centre of the Church-Turing Thesis, taming the wild forces of malware, hackers, and human programming errors.

The Church side defines scientific symbols as *functional expressions*. In contrast, the implementation side provides the confidential data representing the values bound to the computational thread, a computational instance then calculated by Turing's α-machine. The symbols communicate with the object-oriented machine code by substituting the abstract and the physical implementation as one λ-calculus variable. Each digital capability key is an immutable digital pointer local to a Namespace.

The data characterises the object-oriented machine code by matching each symbol within a Namespace context at the same place and time. The digital objects encapsulated by Capability Limited Addressing are media-tight. Media-tight means both function-tight and data-tight. Any atomic service, like (a + b = c), is defined by a single, fail-safe, object-oriented instruction using a Capability Key to represent each λ-calculus symbol in the statement. The technology eliminates malware attacks because no room exists for a sneak attack by malware or hackers. No blind assumptions exist to hide or gain an unfair advantage.

THE PERIMETER OF COMPUTATION

With the arrival of microprocessors, the software genie left the bottle, and the helpful Dr. Jekyll morphed into the killer Edward Hyde. In a network, deliberately coded malware, unfair computation, and communication privileges exist, attacked immediately in von Neumann's unguarded, blindly-trusted architecture. These attacks will only end when the media-tight boundaries of Capability Limited Addressing encapsulate each λ-calculus Variable, validated as a functional expression, protected and preserved as individual computational objects in a network.

Symbol by symbol and instruction by instruction, the immutable access rights of Capability Limited Addressing define a universal, fully protected model of networked computations. The symbols and their relationships to one another define frames of access rights for linked (chained) mathematical expressions. Implementing mathematical computations combines and extends the Abacus and the Slide Rule models by programmatically preserving the Babylonian Golden Rule of form matching function. At the same time, privacy and functionality are validated in-depth and verified in detail on a clockwork frontier of computer science.

The General-Purpose Computer distorts the golden rule of nature that always matches form to function. Natural selection guarantees functional stability through graceful evolution. From the finches on the Galapagos Islands that inspired Darwin and Jacob Bronowski's coiled spring hidden inside the form-and- function of a Rose, or the Abacus, it is balancing the forces of nature in the universal model of computation, using a private DNA that reigns supreme.

When the rampant commercialism of General-Purpose Computer Science took over, the industry adopted a half-baked solution. Its incomplete solution cannot survive, artificially sustained by financial and vested interests to retain a market grip on clients through endless backward compatibility. Thus, the industry ignores innovation to preserve outdated hardware and unreliable software, sheltering the supplier from competition but not from crime. The lack of competitive progress is overtaken and disrupted by better malware and ever more devious hackers, soon to be overrun by the inhuman threats of Artificially Intelligent Malware.

Petrified during the pre-electronic age of the Cold War, the architecture of the General-Purpose Computer has not evolved to the changed environment of global networking. It is increasingly unsafe, hostile, and an eternally expanding international network. The old architecture of the first generation of mainframe computers designed for batch processing unavoidably fails in the electronic village of the 21st century.

The outdated virtual machine was only true-to-form for the single function of a batch process after WWII. It grew into timesharing that prospered as a business model until circa 1980. Since then, bugs loiter undetected, and malware migrates across cyberspace. Corruption reaches across unmarked computational boundaries to silently spy, steal, and disrupt software while remaining hidden. For digital security, digital boundaries are vital. Data-tight software and media-tight functions must operate reliably by remaining true to form.

Natural selection is nature's way of rejecting flaws, leaving the best solution as an evolutionary thread across changing times. Natural selection always follows the golden rule. Physically adapting to the changing environment is how species survive in a hostile, dynamic world. It is how the Abacus began in Babylon, replacing stone piles. Species survive because other alternatives fail, are discarded, rejected, or die out. Natural selection perfected the Abacus to last and adapted it to best fit every culture. The software in a General-Purpose Computer lacks the atomic ability to adapt and evolve gracefully without traumatic risks. Slow progress in recommissioning upgraded software demonstrates this problem.

For example, the Boeing 737 Max, grounded for a year due to software problems, will never achieve the expected commercial success, and Boeing may fall from leadership in commercial aircraft. The same problem threatens every high-technology industry and industrial nation. The centralised design and binary compilations only lead to mistakes, stagnation and digital dictatorship.

Using a Church-Turing Machine that precisely fits the implementation to match mathematics or functions avoids the *'too-big-to-test'* problem that is systemic to General-Purpose Computer Science. A protected λ-calculus computation matches the implementation to one scientific expression, verified and validated one symbol at a time. Likewise, the Abacus performs addition and subtraction by matching the framework to precisely follow the rules learned from the symbols on a chalkboard at school. The dynamically bound modularity engineered by the λ- calculus is the scientific architecture for stable evolution in endlessly changing hostile environments of global Cyberspace.

SYMBOLIC DECOMPOSITION

The expressions of symbols define a limited whitelist of approved computational objects that scope and bind functional mathematics. The atomic symbols for addition and subtraction are just a handful ($a, b, +, -, =,$ and c). Its whitelist is expressed digitally by Capability Keys held in a C-list as a computational frame for Turing's α-machine. Each Capability Key represents one individual symbol, while each C-List represents one expression. Symbolism exists on both sides of the Church-Turing Thesis. One side, the academic side, is the media that defines the symbol scientifically. In contrast, the other side is the real world, where physical objects implement the geometry to scope the design held in memory using symbols that shape each functional calculation.

When Alan Turing conceived his single-tape Turing α- machine, all these symbols were encapsulated and protected logically and physically on a private paper tape. As Turing's α- machine, the result equates directly to Alonzo Church's *'application of variables to a function abstraction'* framing a mechanical form for the functional computation as in Church's 'λ-calculus Application.' Turing's single tape α-machine does not stand alone but is unequivocally at the dead centre of the Church-Turing Thesis, encapsulated by the symbols of the λ- calculus.

Thus, the Church-Turing Thesis perfectly matches the form and function of each idea; one side is abstract logic, not just mathematical, represented as the expression of symbols on a chalkboard, while the other is physical, geometric-digital objects stored in the computer network. Thus framing Turing's *'single- tape'* α-machine as the λ-calculus engine of a 'λ-calculus Application.' It is the scientific solution that locks out malware and empowers functional programming.

PP250 implemented scoped encapsulation atomically programmed as object-oriented machine code. Two types of machine instructions create Object-oriented machine code. Both use capability-limited addressing to identify variables. One type is functional. These are six new Church Instructions. The second type is data, which consists of the various instructions in a RISC machine.

A function abstraction is a digital black box guarded by Capability Limited Addressing. An internal list of Capability Keys defines the scope of the abstract functions. The function abstractions are nodes in a directional DNA hierarchy, defining the DNA of the namespace as a functional evolving digital species. A directional DNA hierarchy of immutable capability keys protects the application's functional behaviour as a digital species. Every thread of execution navigates the approved DNA. At the same time, malware is locked out as unapproved potential terrorists and not allowed to enter the function abstraction, the thread, or the namespace. The directional DNA hierarchy is a nuclear structure of individual capability chains validated and verified atomically, instruction by instruction.

When every machine instruction, including the new Church Instructions, obeys the Babylonian Golden Rule, Babbage's *Infallible Automation* is an achievable result. Every function, expressed correctly and exhaustively tested, is checked in real- time by the guard rails of capability-limited computation. In this context, Turing's α-machine is a programmable digital gene. The gene is a software survival machine that can only track approved chains of protected DNA. The Turing Commands and the more functional Church Instructions stay within the guard rails of the digital context, defined by the λ-calculus safelist of authorised symbols published by each node in the directional DNA hierarchy of Capability Keys. These hardened software-defined objects prevent both software break-out and malware break-ins.

In this pure scientific form, a computer is a Church-Turing Machine, founded on the λ-calculus, positioned at the dead centre of the Church-Turing Thesis. A λ-calculus meta-machine drives the clockworks of Turing's α-machine to achieve the same flawless results that Charles Babbage and the other successful survivable solutions.

THE NAMESPACE CHALKBOARD

The infallible lessons first learned from a chalkboard at school are ideally automated to solve scientific or artistic problems in society fully. The framework of a computational digital gene is not only data-tight and media-tight. It is, therefore, fail-safe for all and any media. It is also program-controlled without unfair default and static privileges that disrupt a General-Purpose Computer. Instead, Bronowski's taut spring is the object-oriented machine code of the encapsulated function abstractions.

An encapsulated, single-tape Turing α-machine model implements the universal model of computation and simple, understandable chalkboard functions program each independent λ-calculus namespace. The CTM navigates one computational thread through the directional DNA hierarchy of symbols. The named symbols from the chalkboard carry variables to the function abstractions as defined by the DNA of the species of an application. The functional Church Instructions[37] not found in a General-Purpose Computer apply the rules of the λ-calculus to substitute variables for each symbol. The Namespace frames the object-oriented functions as a coiled DNA spring following the universal model of computation found in every form of vibrant life in nature.

The chalkboard expressions of symbols form into chained DNA strings, forming the coiled springs of the object-oriented machine

---

37   A functional Church-Instruction dynamically unfolds and binds the access rights to use a functional symbol using from a Capability-Key that extends the context of a computation executed by Turing's α-machine.

code. Just as petals unfold in a Rose, a handful of functional Church Instructions added to the imperative commands[38] of Turing unlock and unleash the coiled springs of the λ-calculus. Functionally, symbolically, orderly, case by case, abstraction by abstraction, boundary by boundary, type by type, and species by species, digital life unfolds, operates, and instinctively survives in a hostile digital world.

---

38   An imperative Turing command is a commands or requests, includes the giving of prohibition or permission. The attributes are all statically bound. An example in English is "Please sit."

FROM ABSTRACTION TO REALITY

Automatically, the clockwork mechanics of Capability Limited computations rigorously check each instruction in a cycle as an indestructible machine instead of loose parts. The mechanisms are programmable and built for functional reliability. The mechanics of this control mechanism are like the chalkboard examples learned at school, which anyone can understand.

A quick overview sets the stage using the Abacus as an arithmetic machine as the prime example. However, a Slide Rule and Ada's program for Bernoulli Numbers and other telecommunication examples used by the PP250 are also referenced. The PP250 details explain the architecture of a Church-Turing computer, used repeatedly simply because no other example exists that is uncompromised by the unsafe centralization of the von Neumann Architecture.

Remember that software in a Church-Turing Machine is a continuous cycle of replication and regeneration as a living species. The process of functional expressions shapes a dynamically reproducible software application with nuclear instances of DNA strings that obey the programmed expressions on a virtual chalkboard. The individual threads in a species exist as private λ-calculus Namespace computations, protected as a nuclear structure of functional mathematics and confidential data, hidden from outside attack. Even close relatives of the same species cannot access confidential data in another thread without permission granted through a Capability lawfully delegated by the owner.

As learned in Babylon and taught at school, computations are simplified when reduced to their atomic algebra of symbols.

Mathematical power now exists incrementally. The algebra of scientifically defined mathematical symbols is the atomic computational structure of the private DNA of λ-calculus objects in a Namespace. The Abacus and the Slide Rule are atomic examples of two private λ-calculus function abstractions. Each has a private DNA of immutable tokens implemented as digital keys in chained lists.

Consider the function abstraction programmed by Ada Lovelace. She programmed mathematically to solve the numbers in a series of Bernoulli Numbers (discussed again in Chapter 4). It, too, is a symbolically defined computation. The abacus is the same but as a functional black box only for addition or subtraction. The tokens on one side of the Church-Turing Thesis are mathematical or functional symbols in expressions.

On the other hand, tokens are mechanised using capability keys with limited Access Rights. The only Access Right to a black box is the permission to 'Enter.' 'Enter' is a frame-changing action performed by the 'Call' Church Instruction. It changes the frame of Object-Oriented Machine Code by reloading the nodal C-List of Capability Keys and the functional program (the single tape), thus moving Turing's α-machine onto the following link in the DNA chain.

An Active Frame defines Turing's (single tape) α-machine. It represents all the private variables (a, b, +, -, =, and c) to and from the links in the chain of object-oriented function abstractions.

{Enter.Abacus}.{Execute.Add} ({ReadData.a},{ReadData.b}) →	This hidden Enter Only

---

Functional Church Instructions can *Enter* an Abstraction.

{Enter.Abacus}.{Execute.Subtract} ({ReadData.a,{ReadData.b}}) →

C-List defines the Object-Oriented Black-Box for an Abacus.

= {WriteData.c}

One Frame is a link in a chain as a context in a Thread defined by a λ-calculus Namespace. The = symbol stands for the Church Instructions Call and Return with the Capability Keys to limit and control the scope of the computation.

FIGURE 1: BLACK BOX OBJECT-ORIENTED MACHINE CODE IN A CHURCH-TURING MACHINE

In Figure 1 and throughout this book, a token as a Capability Key is enclosed with the access rights of the token in a curly brace, for example, *{Enter.Abacus}* or *{WriteData.c}*. Simple arithmetic has one function for addition, shown on the top left of Figure 1 above, using *a, +,* and *b* tokens and another for subtraction, shown at the bottom, using *a, -,* and *b* tokens. The same expressions are found on a schoolroom chalkboard, as exemplified in the bottom line of Figure 2, which outlines the clockwork mechanics of a Church-Turing Machine.

Note that the function {Execute.Add} and {Execute.Subtract} are dynamically bound by the Call instruction using other Capability Keys with the same Load-Capability Key mechanisms. Access ownership limited by Capability Keys defines the computational scope and authority for Turing's α-machine. A Capability Key also references the C-List of Capability Keys, for example *{Enter.aBlackBox}—a* function like *{Execute.Add}* references a block of binary machine code, and the instance variables *a, b* and *c* in the DNA framework are also binary objects constrained as *{Read.a}*, *{Read.b}* and *{Write.c}* Capability Keys.

Thus, a programmed abstraction is protected in depth and detail by an *{Enter.aBlackBox}* Capability Key to the nucleus C-List storing the attendant digital mechanisms for the required sub- structure and any extended DNA to other structures. Theoretically, the Abacus example has two executable machine code methods, while the DNA chain determines the maximum number. One executable method is for addition, and the other is for subtraction. The private capability list for each frame in an extended chain opens an *{Enter.someNode}* Capability Key lists the next step from the chain selected from the DNA hierarchy of the Namespace.

Note that in-scope is different from in-directly-accessable-memory. Thus, the DNA structure of Capability Keys with type-limited access rights to the objects is a blueprint that combines digital security with functional methods but leaves memory management and function distribution as different problems. In other words, cyber security is automatically built into each mathematical expression while leaving the physical memory location for lazy evaluation as and when used.

For example, *Call.{Enter.Abacus.{Execute.Add}}* is a functional Church Instruction using two Capability Keys—the first, *{Enter. Abacus()}* to the abacus C-List and the second *{Execute.Add}* to the authorized method Add() found in the Abacus' C-List. The substituted variables used by the function abstraction are found in a chosen program-controlled capability register of the Church-Turing machine. The Capability Keys must be in-scope, but only the objects needed are in locally accessible memory. It reduces the footprint of the execution to the path chosen. If only the used method is Add(), then the Subtract() method can remain at rest in a home location.

Note that an executed function, like Add() or Subtract(), is substituted on use by the trap and binding mechanisms of Capability Limited Addressing just as any other λ-calculus variable. It is a unique and vital characteristic of the λ-calculus and significantly different from a statically compiled program in a General-Purpose Computer. Using symbols for λ-calculus Variables builds and manages complexity as a hierarchical DNA structure, but the execution is a simple, single path through the structure.

Objects defined in the directional DNA hierarchy are represented symbolically by immutable Capability Keys. Each one has a limited access right in the immutable form *{accessRight.CapabilityID}*. Thus, the imperative Turing- Commands are dynamically bound, verified, and confirmed against limited, scoped access rights. It differs from a statically bound physical address used by an unguarded but shared physical memory page. When form matches function, security is aligned to each object as a symbol in an expression that directly matches the digital implementation. Thus, it can be verified and confirmed on the digital frontier in real time, pipelined with each instruction execution.

Validation and verification occur in depth and detail at the sub-atomic level of direct memory access. For example *{Read.someData}, {Write.someData}* or *{Execute.someCode}* as binary data, used by typical Turing-Commands. Further, Capability Keys held in private nuclear C-Lists are not binary data and only allow *{Load.someKey}, {Save.someKey}* or *{Enter.someClass}*. The Church Instructions (Load, Save, Call, Return, Change and Switch) work with one or more Capability Keys. See Figure 28 for more details. Thus, the PP250 segregated Capability Keys in nodal C-Lists, as first outlined by Jack Dennis[39].

Conceptually and physically, a C-List defines an atomic mathematical expression where the digital form matches the functions of each symbol. Capability Keys to (both directly accessible binary data and indirectly accessible C-Lists) structure the DNA to chain the Object-Oriented Machine Code as linked but independent frames. The C-List stores the DNA string of the nuclear atom, symbolically as an immutable tokenised relationship. Binary data is the sub-atomic structure for traditional read, write and execute objects, the active C-List is the nuclear core defined by an *{Enter.C-List}* Capability Key and addition *{Enter.C-List}* Capability Keys express connectivity among and between functional black-boxes in the DNA chains.

There is no need for a compiled image because abstractions are functionally independent. Functions imported (downloaded) on demand are object-oriented machine code and work the same way

---

39   The idea of Capability Addressing using Capability-Keys stored in C-Lists was published in the 1965 MIT paper 'Programming Semantics for Multiprogrammed Computations' by Jack B. Dennis and Earl C. Van Horn.

browsing works without the downside of malware or hacking threats. The methods used by each nuclear atom are either for public or private functions. The object-oriented machine code of a capability-limited Church-Turing Machine implements all this.

An Abacus breaks down decimal arithmetic into individual decimal function abstractions processed as an array. Its numeric array shares a common decimal function abstraction using substituted variables. Each beaded rail of the Abacus stores the Thread variables, while a human head stores the abstraction procedures for add, subtract and reset. The decimal digits as *'substituted variables'* are named and addressed symbolically, as ordered by a power of ten.

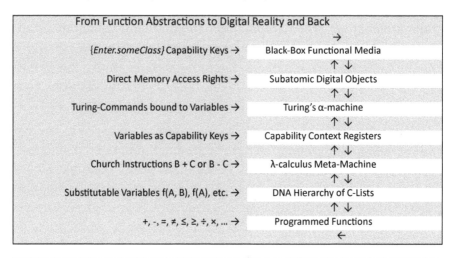

From Function Abstractions to Digital Reality and Back	
	→
{*Enter.someClass*} Capability Keys →	Black-Box Functional Media
	↑ ↓
Direct Memory Access Rights →	Subatomic Digital Objects
	↑ ↓
Turing-Commands bound to Variables →	Turing's α-machine
	↑ ↓
Variables as Capability Keys →	Capability Context Registers
	↑ ↓
Church Instructions B + C or B - C →	λ-calculus Meta-Machine
	↑ ↓
Substitutable Variables f(A, B), f(A), etc. →	DNA Hierarchy of C-Lists
	↑ ↓
+, -, =, ≠, ≤, ≥, ÷, ×, ... →	Programmed Functions
	←

FIGURE 2: THE HIDDEN SPRINGS IN A λ-CALCULUS ABSTRACTION

The secure names of the immutable Capability Keys add a high-level syntax to the object-oriented machine code. The names define the context of computation as the Capability Keys bind digital objects to Turing's α-machine using the universal model of computation. Substitution occurs by saving the local variable as a Capability Key in a C-List or a Capability Context Register.

The Capability Keys serve the same purpose as the fixed wooden frameworks of the Abacus and the Slide Rule. They act to keep the dynamic movement of the specific calculations on high-speed customised rails. The directional DNA hierarchy of Capability Keys defines an end-to-end computational structure. It is a lightweight digital structure of mathematical types obeying the Babylonian Golden Rule. Thus, the digital form matches the mathematical functions. In-depth and in-detail Capability Limited Addressing guarantees valid authorised access rights to the digital components in cyberspace.

At the same time, the imperative Turing commands are no longer bound to static virtual memory. Instead, each instruction names the symbol to which it relates and is securely bound. In this world where Capability Keys equate to λ-calculus Variables, the mathematical expressions on a chalkboard, such as $x \mapsto x^2$, or $fn\ x => x*x$, are written in machine code as a single, fully defined Church Instruction using Capability Keys to stand for each variable. The essential need for a specialised language compiler, a centralised operating system, page-based virtual memory, and unguarded default privileges disappear.

Using functional Church Instructions, explicit Capability Keys substitute functions and variables to allow identical code in different computational threads. The λ-calculus form of $\lambda x.x^2$ binds a thread with the substituted implementations to the substituted method of the function abstraction. Each symbol directly references the digital implementation, bound to Turing's α-machine by a set of capability context registers. A Capability Key unlocks the access rights and binds the digital media to Turing's α-machine using a Church Instruction.

The PP250 provided a C-List for each function abstraction with *Enter* access right as a semi-permeable membrane between domains of secure functionality. When an {*Enter.C-List*} Capability Key is unlocked, a new frame comes into computational focus and takes control of Turing's α-machine. Using the Call Church Instruction unfolds the new framework and saves the return context on the Thread stack belonging to the Thread as a universal model of computation.

Thus, the *Enter* access right prevents interference between nodes by hiding the C-List from inspection until control by the referenced abstraction. Each domain, however small or big it gets, remains a functional '*black box*' hidden behind the *Enter* Access Right.

Thus, a chain of protected functional nodes is linked by Capability Keys, forming the directional DNA hierarchy as a functional tree with a trunk, branches, twigs, and leaves. The tree grows by creating leaves, twigs, and branches that might later die back as computations progress in parallel throughout a network.

Navigating the universal model of computation requires Capability Keys to be inserted into a Church Instruction to unlock the next step. By implementing the universal model of computation, each digital thread is in another function abstraction that includes a private LIFO (Last in, First Out) stack with storage for every machine register.

COMPUTATIONAL THREADS

Threads compute independently, both locally and remotely, across a network. They are self-standing digital genes, software survival machines working as parallel transactions. A Thread performs a Turing-compliant computation on a journey of calculations through a λ-calculus Namespace. Real-world events are bound to a thread through event variables. The survival instinct of the Namespace is in the error-detecting design of Capability Limited Addressing that adds digital girders to the binary fabric. The survival instinct triggered by a scope violation detected by the capability meta-machine underpins and encapsulates Turing's α-machine within Capability Limited Addressing, see Figure 7.

Each resident Capability Key encapsulates a digital implementation created during the Namespace's incremental assembly and application assembly process. In a Church-Turing Machine, the resident symbols as λ-calculus Variables are independent, immutable, and transparent. It transparency controls users' confidential data and their choice of services in their protected Namespace.

Using a private Abacus, the mechanical framework of beads and wires serves the same purpose. Still, advantageously, in a Church-Turing Machine, the immutable Capability Keys form and thus protect the dynamic, programmable framework. This programmable framework targets and retargets the context of execution, frame by frame, as an Abacus for addition or subtraction, as a Slide Rule for logarithmic functions, or as Ada's Bernoulli Numbers, or anything else without any scientific security loss.

43

The ability to target the computational context does not exist in a General-Purpose Computer. Still, in a Church-Turing Machine, two complementary instruction types exist that counterbalance activity to keep the software on mathematical rails, like the Abacus. Functional Church Instructions use the immutable Capability Keys for navigation, leaving the imperative Turing Commands to manipulate binary data as a simple encapsulated procedure.

An *Enter* Capability Key to a nuclear C-List of a λ-calculus function abstraction is a safelist of need-to-know objects. The C- List is the DNA string for the node. At the dead centre of the Church-Turing Machine, the C-List defines both the functions and the form of the functions, preserving the Babylonian Golden Rule. Dynamic access procedures guard the entry gate, the Ishtar Gate, and this digitally walled and defended computational domain. The guard procedures unfold the DNA string needed to process the function and variables provided by the Thread.

The directional DNA hierarchy is trivial in an Abacus. Still, for the global telecommunications applications addressed by PP250, thousands of nodes and tens of thousands of threads link functional chains of reliable computation together dynamically based on calculated results.

These chained computations stay on track, fully restrained by the boundary guard rails plotted by the directional DNA hierarchy of the Namespace as a functional assembly. The symbols of each abstraction define the available scope for computation as each thread progresses. Each Capability Key is a functional guard rail.

Navigating between function abstractions requires an *Enter* Capability Key inserted as a functional Call executed by the thread as a Church Instruction. These Call Instructions unlock the digital gates to the following link in the DNA chain of the λ- calculus namespace. One by one, the immutable Capability Keys unfold the approved access rights of the computational tokens that constrain Turing's α-machine to a single symbol or expression as a private frame for a single calculation.

The immutable Capability Keys are the digital representations for access to functionally limited mathematical symbols. The coiled spring of a Turing procedure is bound to a symbol cached in a capability context register. The context registers limit Turing's α-machine to a single algorithm and a finite set of required variables. One frame is unfolded at a time, in sequence, dynamically on demand.

This generic, multi-function formula was rediscovered by William Oughtred[40], circa 1622, for his Slide-Rule. The scales of each slide mechanically encapsulate the functions of Slide Rules. They are the internal procedures of a logarithmic class of abstraction. In a Slide Rule, each algorithmic calculation is a specific logarithmic scale embossed on sliding rods. The class supports one function selected on demand.

Babbage, with his two engines, engineered his scales as studs on rotating drums and went the extra step to automate the movement of a printer to remove the human errors of typesetting and any unfair privileges. Babbage worked on printing flawless tables to fourteen decimal places. Ada Lovelace[41] went far further. Complementing Babbage, she cemented her extended vision of computer science. Ada recognised the unlimited functional power of the Analytical Engine as a universal machine for programmed functions, including writing, music and art.

She demonstrated her vision as a mathematical program for the Bernoulli series. It survives intact and remains true to form now and forever. Later, circa 1936, Alonzo Church expressed her vision scientifically. Moreover, his concept of a *'λ-calculus Application'* guided his doctoral research student, Alan Turing, to the single tape α-machine.

---

40   William Oughtred 1574 – 1660 mathematician and clergyman circa 1622 used two sliding scales to perform multiplication and division.

41   Augusta Ada King, Countess of Lovelace daughter of the poet Lord Byron (1815 –1852) was an English mathematician, best known for her work on Charles Babbage's Analytical Engine. As 'the enchantress of numbers', she recognized his machine had applications beyond mathematical calculation, and published the first software algorithm ever composed for a computer.

THE HIDDEN SPRINGS

The λ-calculus is the hidden scientific formula behind nature's universal model of computation that Bronowski voiced as nature's *'coiled spring.'* Turing's computational gene with Church's mathematical DNA creates evolvable, software-defined species that exist atomically in a protected λ-calculus namespace. The same coiled spring in nature that makes flowers grow in the warm sunshine unfolds the internal algorithm, distributed by the DNA string to each petal as stable function abstractions. Each petal has a matching physical and functional form—dynamically substituted variables define each petal's character, shape, and colours. Likewise, the universal model of computation secures the leaves and flowers of a software species using Capability Keys as DNA strings for every frame of functional calculation.

A λ-calculus Namespace controls the software-defined petals to grow with assorted colours, specialised shapes, and a variety of sizes, all determined by the combination of worldly events with pre-programmed DNA strings as function abstractions, using the substituted λ-calculus variables with each unique Thread. The formula allows algorithms to pass between frames for calculation, all protected by Turing's α-machine as a digital gene. A Thread defines the status of one instance of Rose, with petals as attendant Threads organised scientifically.

At every step, the digital gene owns the tokens as symbols to access private objects. Shared objects require a conscious action of a delegation that must start with the object's owner, who, depending on trust concerns, can share a proxy object that terminates sharing under chosen circumstances. It keeps data ownership in the hands of a logged-on Namespace owner.

At the same time, all centralised overheads disappear, and no almighty superuser or unfair default privileges exist. Computer science changes from a centralized dictatorial master to a distributed, chosen set of obedient servants. In the fullness of time, cyber society remains democratic, governed through immutable tokens of power as found throughout history. Its power is always in the hands of a cyber citizen and subject to democratic rules.

The digital gene in Turing's α-machine navigates endless, data-tight cellular chains as mathematical black-box frames. Each function is independent, a self-defended, fully protected frame of computation. Like the Church-Turing Thesis, one side of a Church-Turing Machine is a symbolic algebra of functional expressions learned at school (the software, reprogrammable side), composed of pure functional expressions as can be written on a chalkboard.

The physical side stands for the natural world. It is the digital computer side. It uses digital geometry to encapsulate the digital objects that materialise each symbol as data-tight digital media. The assembled symbols define the λ-calculus Variables in the computational namespace as an application, a programmed software species. Each encapsulated function is a reliable, functional part.

Using the Church-Turing Thesis this way, every digital part is an engineered component of a data-tight/function-tight/media- tight machine. Each part has a digital form that matches the application functions, symbol by symbol, where each λ-calculus function abstraction is a class of object-oriented assembler code. Code that obeys the Babylonian golden rule. More than this, symbols are dynamically bound and protected by the object-capability system. The object capabilities deconstruct monolithic virtual machines into the same essential mathematical functions used at school that Charles Babbage chose for his Analytical Engine.

This flawless mathematical architecture comprises true-to-form, protected, functional digital objects. Each frame is program-controlled by the Church Instructions as computational steps with one set of variables, packaged as a digital gene. Turing's α-machine is the survival machine of a computational Thread, with one function in focus at a time. A Thread computes a sequence of digital frames that unfold their coiled springs for each stage as a dynamic species.

The computation framework is pure, functional expressions. Each is encapsulated and executed as an unstretched, unshared, single-tape Turing machine. The transition between frames is program-controlled by equally encapsulated functional Church Instructions. The framework of limited functional genes is computed frame by frame. The Abacus, a slide rule and a Church- Turing machine apply the rule of form-follows-function. They are all cut from this same cloth. The computationally true-to-form digital patterns are physically secure and, like the Abacus or the Slide Rule, are future-safe through pure functionality.

An Abacus and the Slide rule represent two frame types, two function abstractions in the computational video. The video is defined, end-to-end, by symbols and the laws of λ-calculus. The digital video executes a route as a chain determined by a virtual chalkboard of expressions. Each computation thread navigates the directional DNA hierarchy using a private set of event variables.

This true-to-form architecture lasts flawlessly and forever because both mathematics and users are serviced faithfully in full. It is the formula of Industrial Strength Computer Science. A platform fit for future generations trying to survive in a conflicted world, safely sharing the same critical platform through the universal model of computation that allows species to share the hostile world. A Church-Turing Machine has no unfair privileges and no unelected authorities. Frames are always true-to-form; computational boundary rails act like an electric fence detecting hackers and blocking malware on the spot, red-handed if they try to cross an unauthorized digital border, even in the dead of night.

Applying the Golden Rule guarantees that neither theft nor digital corruption takes place without first triggering an immediate, on-the-spot reaction. The rails recognise the slightest anomaly and instantly activate an instinctive fail-safe, program- controlled response. The self-protection from the functional locks and Capability Keys enforced by Capability Limited Addressing verifies, validates, confirms, and ensures the modular, object-oriented machine code stays untarnished and true to form.

CHAPTER 3

The Delusion Problem

*"It is far better to grasp the universe as
it really is than to persist in delusion,
however satisfying and reassuring."*

*Carl Sagan, 1934-1996, was an
American astronomer, cosmologist,
astrophysicist, astrobiologist, and
author.*

In 1967 and ever since, the impact of microelectronics and the digital convergence of communications and computers on competitive global society is intriguing, demanding, and foreboding. Sadly, the talisman of digital convergence today is undetected hacking, unpunished cybercrimes, escalating cyber- war, and false information that is leading towards digital dictatorship.

Supply-side monopolies built unfair powers into the General-Purpose Computer. Power tips against the public, favouring centralized monopolies, criminals, and enemies. It works against a free society. Citizens have no control, and the progress of the nation, as a virtualised democratic nation, is doomed to degrade and evolve as a digital dictatorship. Corrupt and centralised General-Purpose

Computer Science cannot sustain a government of, by and for the people. It is vital for power to be in the hands of the public, or the end game will be a national tragedy, the death of the American experiment with constitutional democracy.

Using General-Purpose Computer Science, cybersociety will turn to fragmented digital dictatorships. The uneven, unfair, and dangerously centralized computer platform tilts against democratic society. The centralized privileges of the superuser make this inevitable. Not only is it biased in favour of the suppliers, but it also schizophrenically empowers crime and corruption. Deluded by virtual reality, unexpected mistakes have tragic real-world consequences as virtual-world hallucinations confuse the users.

Already Artificially Intelligent Malware exploits hallucination exploiting flaws in General-Purpose Computer Science. It subverts the digitally converged electronic village of the 21st century. Software is the new weapon of choice for social destruction. A WMD with more power on the global stage than atomic bombs to win the hearts and change the minds and perception of reality, tools like TickTok take charge of reeducation controlled by the CCP like Orwell's Ministry of Truth.

The power to harm permeates General-purpose computer science, proven by attacks like the takedown of Sony Pictures in 2014, Mrs Clinton's presidential email subversion in 2016, and the NSA in 2018. Such attacks, and many more, are disastrous, not just to the attacked organization but to the peace and stability of the nation.

There are other examples. I add the most noteworthy items to my blog[42] as they become known, but my list only scratches the surface of known corruption. The full extent is unknown because detection is missing in a General-Purpose Computer. However, virtualised democracy cannot tolerate any of them, found and unfound, where unfound must include every undiscovered program bug in General-Purpose Computer Science software.

---

42   Found at http://www.sipantic.com see also https://www.cisecurity.org/

None of this was a problem after WWII or even during the Cold War in that outdated, pre-electronic age before the microprocessor changed the world. Now, with a computer in every pocket and everyone interconnected, the consequence is grave.

Project your thoughts into the future and recognise that virtual reality will permeate every corner of society in less than a decade. AI headsets will be everywhere. Consider that drivers in cars will be the exception. The latest aircraft will fly themselves. Brilliant Artificial Intelligence will drive the world through networked software applications. Progress outstrips all human expectations and understanding. Patched upgrades will not work. The new technology of Artificially Intelligent Malware and AI-enabled hackers will undermine society by undermining cyberspace. As virtualization grows, hallucination and illusion become the way of life. Malware and hackers will find new modes of attack. It is unacceptable, an unnecessary risk for the citizens of an endlessly automated future.

The delusion problem means *'False News'* will escalate to a new level, including fraud and forgery in AI-driven virtual reality. The signs are already here through doctored images, fraudulent accounts on social media, and, most concerning, the two-faced video streams called *'deep fakes.'*[43]

For the virtualization of society to work and stand a chance as a free democratic society, the digital infrastructure of the 21st century must be scientifically flawless. Nothing less will limit the problems faced by a virtualised society. It starts with digital privacy and ownership. Data ownership must be tangible on the surface of digital computers. Ownership must extend up through every software abstraction, correctly distributed to individuals. Ownership watermarks must start and propagate from the source of any change to verify and confirm the integrity of digital authorship.

---

43   A Deep Fake Obama warning about 'deep fakes' went viral in 2019. A realistic-looking video viewed by millions shows President Barack Obama fabricated by artificial intelligence technology speaking someone else words. Visit https://www.msnbc.com/hallie-jackson/watch/fake-obama-warning-   about-deep-fakes-goes-viral-1214598723984

Big Brother is already watching us, while unelected monopolies, criminals and international enemies are already making critical decisions that belong to society. A Global Civil War is underway in cyberspace caused by von Neumann's General- Purpose Computer. The fight occurs on the digital frontier, in the General-purpose computer's unscientific gap between hardware and software, good and evil, truth and lies, hackers and malware, and freedom and dictatorship.

For America and Americans, indeed for everyone using cyberspace, Abraham Lincoln's words resonate *'that government of the people, by the people, for the people, shall not perish from the earth.'* When society is ruled end-to-end by software, society must not be dictatorial. However, the architecture of today's outdated computers is just that. It is now a public concern. When the difference between safe and unsound computers decides the nation's future, the chosen technology matters to the public. Dictatorial suppliers cannot sell General-Purpose Computers to an unsuspecting public. It is far worse than selling firearms. The global virtualization process needs public debate. For cyberwars to be won and for democracy to survive, computer science must return to its roots, the Church-Turing Thesis, or the patriotic dead from every war in history will *'have died in vain.'*

THE INTEGRITY PROBLEM

I ndustrial Strength Computer Science is rooted in the Church-Turing Thesis. A fail-safe machine is vital for the *Infallible Automation* of individual lives, for fail-safe transport, factories, offices, utilities, computer networks, human communications, banking, transportation, and the like. Babbage's *Infallible Automation* removes the risks of human error and deliberate sabotage. Using this architecture, the PP250 prevented all sources of catastrophic and undetected software failure, malware break-ins, and software break-outs.

Babbage's rule of *Infallible Automation* means unattended, fail-safe computation without human involvement. Removing the superuser and the central operating system is a prerequisite test. The test of flawless perfection is when science displaces branded confusion, and computers support civilians and children who lack expert skills. Transparent functionality is the language for this universal model of computation. Its programming language survives on chalkboards around the world.

The citizens can engage through pure, transparent, symbolically defined functionality. It is crucial to satisfy the 21st century and achieve democratic cyber societies. At the same time, as a global communication network, cyberspace must be trusted to be functionally fail-safe, not simply hardware-reliable. In particular, point-to-point connections of the λ-calculus block centralized spies and man-in-the-middle attacks. The bottom line is that hardware reliability, without the *Infallible Automation* of distributed software, lacks the integrity needed to gain human trust.

However, software integrity is only guaranteed by the λ- calculus, as a validated and verified machine of fail-safe function abstractions that implement Babbage's *Infallible Automation* of software. It is critical. Not to be ignored. The modular design of qualified, calibrated software components, each with a measured mean time between failure (MTBF), is a unique characteristic of a Church-Turing Machine. The software function abstractions of the PP250 achieved decades of fail-safe software reliability with a capability-based multiprocessor that calibrated the MTBF for every object-oriented component. The object- oriented machine code detected errors on the surface of computer science following the laws of the λ-calculus. Involved function abstractions are updated as errors are detected.

Well-engineered by the λ-calculus, Capability Limited Addressing detects every software error on the spot, not just address-range errors but all the many types of abstraction errors. Nothing festers and grows into a catastrophe, not only from malware and hacking but also from bugs and natural breakdowns, like those that brought down the 737 MAX[44].

A Church-Turing Machine automates error recovery instinctively as a fail-safe computer reaction, making the software flawless while keeping the functionality in line and under civilian control. Life-supporting computer software must always remain a loyal, faithful servant, not a dictatorial, capricious master.

However, in this world of False News, reliability alone is insufficient. The mantra for Industrial Strength Computer Science, circa 1970, used for PP250, was *'no single hard or soft error can go undetected*

---

44   The software in the 737 MAX fought with the piolts for control over the doomed aircraft. In a Church-Turing Machine the software through the directional DNA hierarchy of Capability-Keys is always under client control.

or cause any digital loss.' An additional requirement concerns AI and deep-fakes. Information pedigree must be available to users, so; *'the integrity of information origination must be verified and validated on demand.'*

THE DEEP FAKE PROBLEM

Detecting a deep fake requires the source creator of the digital information to be known whenever referencing suspect data. The meta-data exists using capabilities allowing ownership cross-checks for legitimacy. Not only must every instruction be crosschecked by the golden rule, but the pedigree of data must also be traceable. Deep fakes originate on the digital frontier where Capability Keys as ownership identities can resolve this problem. Because data ownership exists with capability-limited addressing and the universal model of computation, data integrity exists from an abstraction's DNA. The providence of information protects users from deep fakes. The DNA providence traces back to a verified, watermarked source.

Industrial Strength Computer Science requires solutions to all these issues. They go together to achieve decades of software reliability by preventing hacking, malware, and every form of break-in and break-out, removing centralized privileges, and anchoring human requirements to the computational surface of computer science. The watermark option built into Capability Keys identifies the data source whenever a 'Write' operation changes the data.

Only through the guarantee of the information quality (through software reliability) can a digital watermark with infallible integrity address the information pedigree problem. Integrity applies to every data form, including a point-to-point encrypted communications link. Its point-to-point mechanism avoids any privileged third-party operating system. Thus, digital integrity is engineered functionally and improves over time to the degree society needs.

The government's spy industry will hate these changes, but they are necessary to protect democracy. There is no value in Homeland Security if it destroys democracy. Preserving the foundations for democracy to blossom internationally instead of dictatorships is the top priority.

The PP250 set a 50-year MTBF benchmark for software reliability by improving hardware reliability through a fault-tolerant multiprocessing platform[45] similar to the multicore microprocessor popular today. Industrial Strength Computer Science must exceed current technology standards using hardware redundancy to last for at least one generation. The lesson of Babbage's *Infallible Automation* is removing every source of human interference. Purging the super-user and the centralised operating system is the place to start. It will lead to flawless results and less effort. The skills are simplified when the schoolroom simplicity of the λ-calculus replaces branded designs because everyone good at school can become an expert.

Automating the λ-calculus through Capability Limited Addressing applies all rules of integrity through a clockwork meta-machine confirming and verifying every action of each machine instruction. Thus, the Babylonian Golden Rule tames every wild force to the perimeter of an explicitly bound functional object. The laws of science tame malware and hackers and track the pedigree of data back to a source. As Ada envisioned, endlessly *Infallible Automation* has, in her own words:

'a yet deeper interest for the human race, when it is remembered that this science constitutes the language through which alone, we can adequately express the great facts of the natural world.'

The fail-safe mathematical machine exists inside a seamless fault-tolerant computer network composed of Church-Turing Machines.

---

45   Fault Resistance and Recovery within System 250. ICCC USA 10.1972, Fault Tolerant Multiprocessor Design for Real Time Control. Computer Design 12.1973, Multiprocessor Controlled Switching Systems. CNET/CNRS France 7.1975, A System for the Implementation of Privacy and Security. ICCC 9.1976 Capability Based Systems. Symposium on Operating Systems 11.1977.

It creates an atomically programmed global software platform that frees all functions from undetected mischief. The flawless rules of the λ-calculus define global cyberspace by adopting and embedding the Church-Turing Thesis as nature's universal computational model.

THE TRUST PROBLEM

The universal computation model and capability-limited addressing connect the Church-Turing Thesis through the λ-calculus to the material world. End-to-end, the solution is proven to be reliably trustworthy. A robust Church-Turing Machine is future-safe and fail-safe, universal, and network-ready. There are no arbitrary branded concoctions, centralization, or superusers. Everything interworks through the laws of the λ-calculus as the combined science of Church and Turing.

A λ-calculus meta-machine is the baseline for trusted Industrial Strength Computer Science. Capability Keys and Capability Limited Addressing enforce boundaries following the rules of λ-calculus that bind the λ-calculus Variables, Abstractions, Functions, and Applications together in a trusted Namespace. Its dynamically engineered framework detects errors as real-world events convert into λ-calculus variables that are the nodal frames in the directional DNA hierarchy of a scientific application. A computational thread evaluates the event using the protected digital geometry of the event variables to unlock approved symbols as both media-tight and data-tight objects in an application context of private computations, executing at the dead centre of the Church-Turing Thesis.

A Namespace stores the assembled symbols as a blueprint of a software species. Type-controlled object boundary parameters encapsulate the nuclear-structured functions of the digital media.

The programmer-engineered directional DNA hierarchy maintains order and evolves gracefully through natural selection as better abstractions emerge over time and experience. The hierarchy formed by the Capability Keys explicitly defines every relationship atomically.

Unlocking access rights between objects requires pre- assembled knowledge as a gifted set of capability keys, a dynamic object creation, or a respectful introduction returned in a handshake. The capability keys guide the Church Instruction to unfold other symbols in a pattern defining the digital object's geometric structure, each protected by Type Limited and Capability Limited Addressing. As nature's spring, DNA always returns to safe conditions.

The nodes in this directional DNA hierarchy of symbols, called C-Lists by Jack Dennis, define the function abstractions as media-tight, data-tight, and function-tight digital form[46]. Malware is locked out because no gaps or cracks exist in the framework. As a sequence of digitally framed function abstractions, this computational framework is as dependable as an Abacus, as multi-functional as a slide rule, as visionary as Ada Lovelace, and as flawless as Babbage's incomplete mechanical engines.

Furthermore, the directional DNA hierarchy guarantees all information's providence, pedigree and integrity. A digital watermark starts with a Capability Key in a Namespace with the same certainty of origin as a signed legal document from the pre- electronic age. Digital convergence demands this. Cyber- dependent generations who depend on life-supporting software in a virtualised society expect flawless mathematical perfection as the only way to proceed with certainty. Industrial Strength Computer Science applies both sides of the Church-Turing Thesis to create a future-safe machine where software and society safely coexist in a flawlessly shared, privately managed cyberspace.

---

46   Media-tight is implies a 'water-tight' container but applies to all forms of digital media from simple numeric or alpha numeric strings to programmed functions and function abstraction of all kinds including for example the digital media that encodes stores and plays sound and video.

Scientific measurements verify the level of trust. Data quality must be measured and guaranteed, as with food, air, and water. The MTBF measures the quality of data while preventing corruption. Capability Limited Addressing measures MTBF while detecting and preventing corruption by reporting the time between error events by abstraction. The PP250 focused on the abstractions with the lowest MTBF and quickly achieved application software stability.

Software in the 21st century is like PP250, a non-stop utility, and like electricity, clean water, and uncontaminated food, it must be constantly checked and filtered. The meta-machinery of λ-calculus provides the independent capability to service all the hidden aspects needed to achieve Industrial Strength in Computer Science as a self-sustained, continuous process. If not, society degenerates from the growth of digital pollution caused by cyber wars and unrelenting cybercrimes. Digital pollution unravels trust in all laws, order, justice, and good government. Conflicts flow from the digital battlefields into the material world from fights that envelope cyber society with existential threats. The progress in Artificial Intelligence increases the danger level, and the risk of digital enslavement to super-human malware starts long before reaching the point of singularity[47].

The exposed privilege in every General-Purpose Computer makes breakout easy. If Artificially Intelligent Malware becomes the ultimate dictator, the society of nations will return to branded dictatorial kingdoms and a medieval form of digital serfdom.

Trust in the automated future of cybersociety depends on a level, calibrated playing field, with defended digital perimeters and dynamically guarded gates implemented at the dead centre of the Church-Turing Thesis. The functionality in cyberspace must serve civilisation without distortion, criminal subversion, or unelected

---

47   Singularity is the point at which machine intelligence and human intelligence cross. *The Singularity Is Near: When Humans Transcend Biology* is a 2005 non-fiction book about artificial intelligence and the future by Ray Kurzweil. Kurzweil describes accelerating returns and predicts the exponential increase in technologies like computers, genetics, nanotechnology, robotics and artificial intelligence. Singularity as Kurzweil says build machines infinitely more powerful than all human intelligence combined.

dictatorial privileges. The electronic village cannot tolerate Artificially Intelligent Malware breakout in any form. Harnessing the illusions of virtual reality serves instead of rules society. It is hidden unfair privileges that cause undetectable corruption. It is pure mathematics that is vital to the Virtualization of Democracy. The Church-Turing Thesis is crucial. Understanding this cannot come too soon.

THE PRIVILEGE PROBLEM

General-purpose computers use shared memory with rings of default privilege. The opaque control means privilege is shared unfairly. It only leads to the triumph of evil over good. Equality under the law is missing where and when it matters most, on the surface of computer science as instructions, fire expending energy. Here, even mundane software, even a single machine instruction, can be an unrecoverable, catastrophic mistake. Plessey foresaw this and the inevitable endgame that must prevent corruption and avoid unfair privileges. The architecture for digital convergence must be clean, continuously correct, end- to-end, point-to-point, structured atomically, and scientifically flawless. A computational machine implemented by mathematics. A computer without any threats from default privileges or branded best practices.

Peter Denning asked me to write about PP250 in 1976, but my story was incomplete. Digital convergence had begun; however, timesharing ruled computer science, and the illusion of an all- powerful operating system kernel fascinated and deceived the mainstream. Only outsiders who needed reliable software searched for better alternatives. Homegrown batch processing allowed mainframes to continue using von Neumann's flawed assumptions, upgraded by Cold War preference for top-down, bureaucratic, card-carrying security clearance. It led to using identity-based access control and human best-practice administrators as the security system for mainframes and general-purpose computer science. Software reliability did not matter to corporations building homegrown applications with teams of committed, on-site IT staff to enforce corporate *'best practices.'* After a crash, the next shift just restarted a cleaned software rebuild.

The communications industry needed more. The service is global and continuous, and unattended exchanges need decades of software reliability for the international customer base.

Consequently, the unique design of the PP250 awoke new interest in Capability Limited Addressing, and with the CAP machine developed at Cambridge University, the architecture came to the fore in 1977. At SOSP'77, held at Perdue University, the published results favoured capability-based computers, but then and now, these facts remain ignored[48]. The mainstream was wedded to the dream of the all-powerful operating system kernel solving all problems, including all cybersecurity problems.

The vested interests that dominate the industry overruled any change. Batch processing worked eight-hour shifts between cold restarts, and the persistent IT staff hid all software problems behind locked doors. These branded designs kept their grip on the market, and as von Neumann ignored the λ-calculus, the industry ignored Capability Limited Addressing. Exaggerated claims from kernel supporters drowned out the arguments for converting to industrial-strength Computer Science. Led by Butler Lampson[49], Capability Limited Addressing was dismissed as a 'special case' only needed for telecommunications.

Since then, batch processing ended, and mainframes died out, while Identity Based Access Control and Lampson's operating system dreams failed to deliver cyber security. At the same time,

---

48   My capability presentation can be found at SIPantic blog www.sipantic.com for August 2019. No others on the panel documented their positions.

49   Butler Lampson is a Technical Fellow at Microsoft Corporation and an Adjunct Professor of Computer Science and Electrical Engineering at MIT, inspired the Identity Based Access Control security ideas of the Vista Operating System.

semiconductors, personal computing, Internet browsing, and multi-programmed digital convergence took centre stage. With the arrival of smartphones circa 2010, the 'special case' dismissed by Lampson became the general case.

The outrageous claims of Identity-Based Access Control using Access Control Lists never materialised. When the much-touted Vista Operating system[50] failed in the marketplace, monopoly suppliers quietly gave up on cybersecurity and punted the problem to specialised third-party solutions, as listed from left to right in Figure 8, and discussed in depth at that point.

Cybersecurity remains opaque and baffling as an unsolved problem in general-purpose computer science. The General- Purpose Computer is a catastrophic threat beyond the scale of the Titanic, caused once again by compartmental breakout, missing boundaries, unfair privileges, and unproven practices. The leaks, gaps, and cracks are resident traps that cannot hie; the next generation of intelligent malware bypasses all Byzantine security software and the latest patched upgrades. Like the Titanic, the problem starts below the waterline, flooding the engine room of a General-Purpose Computer. Just patching over a digital mathematical gap or logical cracks in the fragmented, proprietary voids of General-Purpose Computer Science cannot fix the unearned default privileges and shared authority that still is unaddressed. The mainstream was never interested in the systematic solution to cybersecurity. Instead, the responsibility was kicked over to application programs, increasing the need for sophisticated best practices and guaranteeing the problem could not be solved.

A General-Purpose Computer demands constant human attention, which Babbage worked so hard to avoid. Urgently supported, hasty patches and the crucial need for regular software upgrades still cannot

---

50   Vista Operating System, released by Microsoft for consumers on January 30, 2007, was widely criticized by reviewers and users. Due to issues with security features, performance, driver support and product activation.

guarantee the success of cybersecurity. After promising the earth to avoid the truth in 1977, operating system theorists did not deliver. After Vista failed, they walked off the set, leaving cybersecurity in total disarray by circa 2000 with only patched upgrades to work with.

Upgraded patches are late, always after the first new attack succeeds. They offer no hope of preventing the next *'zero-day'* attack because nothing removes the default privileges or the blind trust in the shared memory and the almighty (imperfect human) superuser. A top priority is perfect software and honest, flawless system administrators. General-purpose computers forever depend on unworkable expectations that only offer 'too little too late'. The impending, foreseeable catastrophes require the complete application of proven science. Patched upgrades cannot stop a massive data breach using default superuser privileges from an unknown remote location. Any stolen credentials make this easy.

Stolen credentials become undetected, unpreventable attacks on a General-Purpose Computer. The attack vector exists by default. It is the cause of the long delay to the discovery that reverberates for months and years after each such attack. The Cold War hangover only led to unavoidable over-regulation and stagnation caused by branded dictatorships. Nothing in General- Purpose Computer Science deals with these massive attacks. Nothing provides any warning, and nothing prevents the echoes from the data breach of credentials from reverberating without later detection throughout the world.

The assumption of blind trust in perfect software clashes with the technical innocence of a virtualised society. It only works for experts like von Neumann. The written constitution never expected such inhuman threats to attack the nation. There is no remedy beyond fidelity and the independence of all three government branches, but cyberspace is centralized and lacks diversity. In a network of General-Purpose Computers, malware spreads in a flash, and independence does not exist. These unfair privileges exist as hardware flaws that software cannot fix. The science of λ-calculus and the universal model of computation must take over for the electronic village to survive and for civilization to prosper.

General-Purpose Computer Science is a tragedy waiting to happen. No one can secure the nation's defence; bulk data is too easily stolen, even at the NSA, the CIA, or the FBI. Pure scientific functionality through the Church-Turing Thesis is the only way to level the privileges in cyberspace. Using a Church-Turing Machine that follows the laws of λ-calculus to shine, remove default privileges, and prevent ambient authority sharing in a virtual machine.

The science of the Church-Turing Thesis is the way to apply mathematics smoothly and seamlessly across cyberspace and to remove the danger of default and superuser privileges. These dangerous default powers evaporate by reversing von Neumann's false assumption of blind trust and regulating every case of authority in cyberspace. Using Capability Keys authority is regulated on the digital frontier, as the scientific surface of computer science moves up through the directional DNA hierarchy of a namespace into the hands of the owner of the digital information.

THE DETECTION PROBLEM

P atching does not get the negative attention it deserves. A patch adds software overhead, increases maintenance and support costs, undermines the original architecture, and introduces new best practices. Urgent patches worsen everything without addressing the root cause built into the hardware while creating added threats yet to be discovered. This unstable concoction is treacherous because of the opaque baggage of a failed kernel strategy.

Not even the NSA, CIA, FBI, or other institutions of the US Government can prevent massive data thefts that end up on WikiLeaks or in the press. Software bugs, malware, and hackers stay undetected. Discovery is after the event by other means, and this takes time. Only whistleblowers publicise their work. Enemies of the state keep their successful attacks a secret. They insert malware for later use when a full attack takes place. The patch is too late and misplaced when stolen credentials are applied elsewhere. None of these later attacks fails, while the stolen credentials stay unreported and valid for criminal use without detection. Ironically, using stolen credentials never counts as a cybersecurity failure. In truth, when millions of identities are all stolen at once, it is an adverse event for the institutions of society.

Administrators like Edward Snowden[51] and Bradly Manning stole an entire database and escaped unnoticed until they chose to show

---

51   Edward Joseph Snowden is an American whistle-blower who copied and leaked highly classified information from the National Security Agency in 2013 when he was a Central Intelligence Agency employee and contractor.

their theft. The same was the case at the CIA when a massive data breach of all weaponised malware developed by the US Government was copied and published as Vault 7 on WikiLeaks on 7 March 2017. This enormous data breach stole all the weaponised software designed for use by the US Government to spy and attack others.

This data breach proves how dangerous General-Purpose Computer Science is to virtualised society and the nation. Now, enemies and criminals can use the best weapons developed by the skilled resources of the US Federal Government. Nothing is safe in a General-Purpose Computer; these problems will only grow as the network expands. Beyond all comprehension, the risks will develop over the coming years until the cyber bubble bursts and a catastrophic climatic end occurs. Justice, when possible, is slow, and the most dangerous international criminals easily evade the law.

Such is the case with Edward Snowden and, more recently, the formal accusation of 13 Russian nationals by Robert Mueller's investigation of interference in the 2016 election[52]. The Russians made a *'sustained effort'* to hack emails and computer networks. It began years earlier when, in 2016, John Podesta opened a clickjacking attack[53]. The attack unfolded to steal data and emails throughout Mrs. Clinton's organization. It took years and millions of dollars before the special prosecutor's office issued the (futile) subpoena, and the blame game is still reverberating across American democracy.

This significant example cost millions of dollars and took years to uncover. It is self-evident that unreliable software and the inadequate branded laws of General-Purpose Computer Science threaten the existence of democracy both directly through digital corruption

---

52   On February 16, 2018, Robert Mueller indicted 13 Russian individuals and 3 Russian companies for attempting to trick Americans into consuming Russian propaganda that targeted Democratic nominee Hillary Clinton and later President-elect Donald Trump.

53   In March 2016, the personal Gmail account of John Podesta, a former White House chief of staff and chair of Hillary Clinton's 2016 U.S. presidential campaign, was compromised in a data breach, and her emails, many of which were work-related, were stolen.

and indirectly through government corruption. Society expects and deserves better. Indeed, only the absolute best is enough. The science and technology from the Church- Turing Thesis are vital in the nation's defence and in achieving a citizen's cyber democracy.

The General-Purpose Computer is an unregulated weapon of war, sustained by spy agencies who foot the bill for this anti- social, undemocratic solution. Cars and planes work hard to save every traveller's life; computers must do the same, and the military, industrial triangle, and spy agencies must fall in line. The lack of deep and detailed error detection on the surface of computer science is a mistake that must change. The default powers misappropriated by spies, criminals and enemy interests extend internationally. It is an industrial disgrace, and since the solution has been known and proven for decades, it is ethically unacceptable. Keeping this from public debate is a way for the industry to sustain revenue and avoid loss of monopoly control.

Computer software is only fail-safe when the λ-calculus applies. Using Capability Limited Addressing, error detection occurs at the computation's time and place, at the sharp point when software meets hardware, and one machine code instruction at a time. Living science is engineered and constrained by the universal laws of the λ-calculus as a pure and trusted digital medium end-to-end across cyberspace.

In the meantime, proprietary standards are destroying the future, and worse, the end will accelerate as Artificially Intelligent Malware becomes the weapon of choice for hacking and malware. Inferior quality software, malware, and unfair privileges steal democracy from a virtualised society. It is a foolish nation that dares to depend on General-Purpose Computers in the age of Artificially Intelligent Malware.

THE EVOLUTION PROBLEM

For decades after PP250, neither the Church-Turing Thesis nor λ-calculus received any attention, and the solution remains hidden. Even CHERI[54], Cambridge University's second attempt at using capabilities, remains a hybrid design compromised by shared memory and centralized operating systems from von Neumann's WWII architecture. Cumbersome compilations and ad-hoc patching remain the only options to limit an evolving global threat. Monopoly interests and the interests of the spy agencies override concerns for citizens and the survival of democracy.

Virtualizing society means software must implement law, order, and justice equally for everyone. However, branded monopolies create these laws today. They tilt the power scales to support their continued digitally corrupt dictatorships. A virtualised constitution and government institutions subverted by centralized computer science have no checks and, thus, no balances. It is unacceptable for the everlasting future of society. A cyber democracy, as a government of the people, must be controlled by citizens, just as a constitutional democracy dictates.

The universal chalkboards in schools and universities worldwide express a scientific language of named, meaningful, functional symbols, defining scientific formulas. For the precious global platform

---

54   Capability Hardware Enhanced RISC Instructions (CHERI), a hybrid architecture that build on existed computer systems. See https://search.darpa. mil/?query=CHERI&facet.field=ContentType&facet.field= Topic&facet.field=FileType&facet. field=Offices&page=1&pagesize=10

and an endless future, proprietary conventions are unacceptable. They are self-serving the dictators. They only maintain corrupt power, served by fraud and crime. Then and now, supporting henchmen do the dirty work to line their pockets with gold, as in the Middle Ages.

Citizens must regulate power in cyber democracy. It is too dangerous to do otherwise. Power must be decentralized and distributed incrementally by need as immutable tokens like a chain of office or for cyberspace by a capability. Industrial monopolies, binary conventions, spy agencies, and engineering teams cannot act alone as digital dictators of centralized kingdoms, where corruption and payoffs keep things this way.

Meanwhile, cybercrimes and cyber warfare are becoming the most significant threat to the nation's future. At the same time, the claims made by experts following Butler Lampson in the mainframe age prove false. In centralized computer science, malware and cybercrime only grow. Superuser software cannot crush the problem. They all prove inadequate.

The destiny of the cyber democracy hangs in the balance after draining swamps, engineering dams, and building bridges to last for centuries, then reaching and returning from the Moon. Who can doubt civilizations' ability to use Industrial Strength

Computer Science? Mathematical machines, like the Abacus, the Slide Rule, and Church-Turing Computers, are flawless forever.

Our children's children require this monumental form of Industrial Strength Computer Science—a digital algebra and programmed geometry that protects the smallest atomic detail deep inside cyberspace. Encapsulating software in Turing's α- machine and Alonzo Church's λ-calculus protects each computational frame in the universal model of computation. They solved the problems in depth and detail by putting citizens in control of their confidential data with the programmable security policy of delegated Capability Keys. This secure model starts by reversing von Neumann's flawed assumption.

'Software-cannot-be-trusted' is the assumption to meet, and a λ-calculus meta-machine is the foundation technology for the task.

This program-controlled DNA framework of Capability Keys applies the laws of λ-calculus to modular object-oriented machine code executed as λ-calculus frames by Turing's α- machine, one algorithm, one digital object at a time. Each calculation is a fully protected frame of computation held in a data-tight, function-tight, media-tight encapsulation. Each encapsulation applies the Babylonian Golden Rule of form matching function to regulate software on the surface of a digital computer at the dead centre of computer science.

Both sides of the Church-Turing Thesis are now satisfied. All best practices are automated by decomposing monolithic privileges and then distributed according to mathematical formulas of need to know. Babbage's *Infallible Automation* now serves both science and society simultaneously. All the human tasks are automated, using lock-and-key access rights that double-check power and authority. Most importantly, the keys to confidential functions and critical data ultimately belong to citizens serving a democratic cyber society.

The directional DNA hierarchy runs like Darwin's Finches on the Galápagos Islands. Evolution is stable, graceful, incremental, and environmentally sensitive. There are digital finches in cybersociety. Each one demands specialization for local conditions well outside the monopolistic interests of self-serving corporations. For example, Facebook, Google, and every other digital supplier cannot guarantee user data privacy in virtual machines. This separation is, however, a natural feature of the λ- calculus, the direct consequence of locally substituted variables and downloaded function abstractions as implemented by a Church-Turing Machine.

A Church-Turing Machine controls the security policy through the directional DNA hierarchy of a private Namespace. The security policy of a client namespace is customised independently and, thus, evolves gracefully. The programmed functions are separate from the data and can evolve independently. These two dimensions of ownership are independent and brought together in a thread at run time under the control of the Namespace owner. It is the user's private chalkboard.

THE PRIVACY PROBLEM

C yberspace will expand forever into the next century, and national dependence will grow. Thus, the nation's future depends on cybersecurity. It is the extension of a discussion I began with Michael Clark, the son of Plessey's founder, on a flight from Liverpool to Essex in 1970. Given the growth of weaponised malware and intelligent software, the threat to the nation might seem intractable. However, Babbage's ideal of *Infallible Automation* is a purely scientific problem. A λ-calculus meta- machine is all it takes to enforce the laws of Infallible Automation.

Ignoring this scientific solution means disruptive government regulation will occur, but as with patching, it will not address the root cause. It will suppress innovation and increase crime. Cyberspace must work for every citizen and every corporation, especially those without the resources of highly specialised skills. Indeed, the lack of specialised skills is the root cause of insecurity. A Church-Turing Machine solves the lack of data privacy. Change must begin by ending the curse of shared memory, banning rings of privilege, removing the need for staffing superusers, and separating data for simplified control by the owning users. Cybersecurity must come first through a move to a Church-Turing Machine architecture.

Like everyone else, I began with the Turing machine, and Capability Limited Addressing was the means to a better end. However, this is not the beginning. Like other computers, the PP250 was considered a trade-off between proprietary designs. Few recognised the long-term significance of cyber security and the importance of the λ-calculus. PP250 began by preventing undetected errors in statically shared memory. This decision alone inevitably led to Industrial Strength Computer Science.

Understanding Alonzo's search for the foundations of mathematics and nature's model of computation explained in Richard Dawkins' books, combined, they realize the everlasting science of Church-Turing computations. The Church-Turing Thesis is both the beginning and the end of a search for a safe technology for digital convergence and trusted software as the faithful servant of evolving cyber societies.

Cyber societies must co-exist as intelligent species, but they share human weaknesses. Stable evolution and *Infallible Automation* are vital for this endless future of life with a digital extension. Cyberspace will then embellish, enhance, preserve, and protect human life. Individuality is crucial to empower the progress of civilisation. It is the unfinished chapters in Marshall McLuhan's seminal work on the cornerstones of media theory. He died in 1980 but foretold the media mashup in his first publications decades earlier. His influence dominates the correct understanding of the relationship between humanity, computers, and communications.

Cyberspace is another functional extension of the human form. If wheels extend the legs, computers extend the mind. Protecting the creative individuality and privacy of the human mind is critical to the progress of civilised cyber-society. The success of computer science will be measured not just by breakthroughs using supercomputers but by the ability to educate and preserve the institutions of a democratic society. The digital shadows in cyberspace belong to a real-world owner. Colonizing cyberspace is the task of maintaining and protecting the sanctity and individuality of life. The task is to extend an individual's private DNA into cyberspace safely.

A shared experience is encouraged, but a shared implementation is a threat. Replacing the General-Purpose Computer with Church-Turing Machines is essential to protect individual privacy and the civilised control of democracy. Citizens must own their digital shadows in cyberspace.

A digital shadow is the private extension of an individual into cyberspace. Through fault-tolerant, fail-safe software, the private namespace of the universal model of computation extends nature into cyberspace. While software may not be perfect, it is always private and fail-safe using a Church-Turing Machine. Each Namespace is a secure and private part of a global software machine. It is a safe and correct expression of the owner's DNA and as accurate to form as the related mathematical expression on a university chalkboard.

The individuals are unsullied by shared infections and unharmed by unfair privileges. The resulting *intellectual* privacy is an extension of the human mind that supports cyberspace's continuous and graceful colonization as a society using an irreplaceable form of democratic government. Pure mathematical cybersecurity serves science, individuals, and society equally and forever.

THE BREAKOUT PROBLEM

E qually important, a Church-Turing Machine solves the threat of A.I. software Breakout. The engineered design binds the software to mathematical rails as an algebra of defended geometric digital properties. Capability Limited Addressing detects and prevents every digital bit of the perimeter boundary from external attacks or break-out attempts. Every atomic action in each nuclear relationship is held in place by a string in the DNA chain as a blueprint of authority. The assembly of software builds a Namespace, a namespace populated by the λ-calculus Variables, Abstractions, Functions, and Applications that define everything necessary for software to obey the λ-calculus in each private Namespace.

A λ-calculus namespace replaces the static memory and a complex virtual machine with a private extension, a private digital shadow for everyone, a community, or a nation with a DNA engineered mathematically for the species. A unique directional DNA hierarchy controls the namespace's information and functional security policies. Individuals can subscribe to the groups indirectly through their own private Namespace. Malware cannot break in or out because the λ-calculus variables are mathematically true-to-form. The DNA of symbols protects the atomic and cellular chains as calibrated components. A Namespace is an algebra of scientifically bound constraints for an individual or an organization of individuals.

Digital activity is constrained to a context expressed by the symbols as a class of mathematics defined by the DNA structure of the Capability Keys. Threads compute with one set of event variables at a time. The Thread navigates the DNA using function abstractions to reach a result. This computational form Alonzo Church called a λ-calculus Application. A namespace protects both the directional DNA hierarchy and the confidential data of the λ-calculus Application. No more and no less exists than the symbols, expressions, and privileges as written on the private chalkboard for the namespace.

The λ-calculus meta-machine program controlled by functional Church Instructions applies the laws of the λ-calculus as a universal model of computation. Hacking and malware cannot take place. Unlike a General-Purpose Computer, where security is abstract and calculations are suspect, Industrial Strength Computer Science reverses everything. Computations are abstract, and security is absolute. Further, the user owns and controls the Namespace, not a branded dictatorship.

Enforced digital security prevents breakout. Preventing software breakout is critical in the age of Artificially Intelligent Malware. The MIT cosmologist Max Tegmark, co-founder of the Future of Life Institute, explains the problem of A.I. breakout in his book 'LIFE 3.0, Being Human in the Age of Artificial Intelligence.'

Professor Tegmark explains how superhuman intelligent software, at the point of singularity, will run amok if software containment cannot occur. Using a General-Purpose Computer breakout is easy; Ransomware proves this. Breakout is the forcing function to change to the Church-Turing Machine. A.I. software with 'game-playing' abilities beyond human understanding will alter every calculation concerning cyber dictatorship, cyber war, and national survival.

Even before reaching the point of singularity, Artificially Intelligent Malware will elevate successful attacks on critical resources in cyberspace. When software is more thoughtful and faster than humans, all the best practice expectations of General- Purpose Computer Science will be useless. Preventing super- intelligent breakout is vital to counter this extreme form of malware destroying society. AlphaZero[55] from Deep Mind, a sister company to Google,

---

55   AlphaZero is a computer program developed by the Alphabet-owned A.I. research company DeepMind. On December 5, 2017, when AlphaZero was introduced, within 24 hours, it achieved a superhuman level of play in three different games, Chess,

has proved to be a self-trained software application that wins against the grandmasters in GO, CHESS, and SOGI[56]. After a few hours of learning the rules and playing alone, new winning moves, discovering unknown winning strategies, advanced moves, and new approaches previously hidden throughout human history. In the wrong hands and playing a crazy game of sabotage in cyberspace, humanity and AI will always exploit the flaws in General-Purpose Computers. Human best practices and patched upgrades cannot respond.

Severe consequences are inevitable. One example is a ransomware attack conducted globally by Artificially Intelligent Malware on financial institutions. Blockchains will vanish when machine code encrypts or disables every interworking block or server. An infiltration phase precedes the final invasions, and then, when the sun comes up on a given day, Armageddon will take place. The Future of Life Institute regards cybersecurity as the vital next step for survival before reaching the point of singularity. It is discussed again in the final chapter of this book.

It seems like science fiction, but criminals and enemies will use superhuman software faster than others. Those without scruples will disregard the Future of Life Institute. Super malware is just a variation of the Babbage Conundrum concerning hidden human errors. In Babbage's days, human errors related to data values and data the quality of results calculated to fourteen decimal places. However, in the future, if not already, human interference in electronic villages using artificially intelligent malware and superuser privileges will attack civil society's life- supporting platforms. They will turn the industrial world upside down by wrecking global supply chains on which the unguarded silk roads of cyberspace flow. The damage will break the nation before the scope of threats is appreciated.

Governments, scientists, and engineers must act to prevent this awful scenario. Babbage's idea of *Infallible Automation* must avoid all

---

Go and Shogi.

56   Shogi, also known as Japanese chess or the Game of Generals, is a two- player strategy board game native to Japan in the same family as chess.

forms of his conundrum by transitioning to Church- Turing Machines that prevent any superhuman software breakout. It is urgent. It cannot occur on a supplier's timeframe, wedded to backwards compatibility. The threats of Artificially Intelligent Malware are here today.

Von Neumann's assumptions are outdated, blocking the progress of the civilised world. Computer scientists must change to use Industrial Strength Computer Science. The changeover needs to be hastened by government policy and competitive investment in the Church-Turing Machine. The headlong race to a human-made disaster cannot destroy the nation before the benefits of A.I. software and virtualised democracy gain traction.

CHAPTER 4

The Integrity of a Machine

'To the nations and peoples of every
language, who live in all the earth:
May you prosper greatly!'

King Nebuchadnezzar's Dream of a
Tree, New International Bible, Daniel 4

U nlike the past, the future is defined by, and life is now run by invisible, globally exposed, inherently unreliable, monolithic software. When monolithic software scales in networked cyberspace, law and order fragments. The General-Purpose Computer only scales incorrectly by increasing word lengths, adding centralized memory and growing networked fragmentation. Still, this increased sharing adds new threats, fails to network, and the probability of undetected corruption grows. Hidden bugs and downloaded software destroyed the viability of von Neumann's architecture.

The problem with networking is safe replication and smooth distribution, not a vertical expansion of memory. The universal model of computation scales atomically. Networking a General-Purpose Computer is challenging because of the atomic substance it compiled as a proprietary monolithic construction with arbitrary binary conventions that are impossible to check and do not interwork smoothly or scale gracefully.

It creates a range of *'Confused Deputy'* attacks[57] that exploit the inherent architectural differences and technical problems with interworking monolithic software. The Confused Deputy attack was first named by Norman Hardy when timesharing was popular[58]. The attack exploits malleable binary data to misrepresent the exchange of variables between branded monolithic systems through fraud and forgery. The attack reappears wherever binary data is used on a branded interface to pass critical information between independent software entities.

It is the root cause of 95% of successfully networked malware. The weakness is untyped binary data. Checking binary data is impossible since context is vital for understanding. Checking for forged data or corrupt character strings is impossible. At the critical moment of execution, blind trust takes over, and computer science becomes an oxymoron.

Immutable data called Capability Keys is the only way to prevent fraud and forgery in digital computers. The λ-calculus has no such attack vector. With the λ-calculus, scaling is atomic, using immutable symbolic names impervious to malware. Authority expressed by capabilities follows need-to-know principles expressed by systematic functionality instead of proprietary convenience, thereby preventing all forms of corrupt exchange.

The Abacus mechanically demonstrates visually how the λ-calculus scales atomically using point-to-point interactions—a natural

---

57   A confused deputy attack has several forms that confirm the need for capability-based security. When a virtual machine with superuser privileges is tricked into using this authority it always leads to disastrous consequences. The attack escalates the attacker's privileges by passing forged or fraudulent data and can be equally well performs locally or over a network as a click-jack attack. Identity Based Access Control and Access control list-based systems cannot detect or prevent these attacks. Norman Hardy, The Confused Deputy: (or why capabilities might have been invented), ACM SIGOPS Operating Systems Review, Volume 22, Issue 4 (October 1988).

58   Norman Hardy, The Confused Deputy: (or why capabilities might have been invented), ACM SIGOPS Operating Systems Review, Volume 22, Issue 4 (October 1988). A confused deputy is a program innocently fooled by another to misusing their authority. It escalates the privilege of the attack with the ultimate objective of gaining superuser authority and seizing control of the computer and the operating system. See https://en.wikipedia.org/wiki/Confused_deputy_problem

peer-to-peer solution of authorised relationships without any corrupt mechanical privileges. Monolithic software cannot scale atomically as individually protected function abstractions. The unfair privileges and the centralised digital dictatorships tilt General-Purpose Computers downhill by causing digital chaos.

The λ-calculus is not fragmented, bartering power (Capability Keys) just like money. The architecture scales atomically as a scientifically level playing field. The result is equality and fairness as Industrial Strength Computer Science applies Babbage's hallmark of *Infallible Automation* to prevent the Babbage Conundrum. The digital clockworks of a λ-calculus meta-machine guarantee pure mathematical function abstractions are bound together correctly, fraud and forgery-free. Atomic software, when bound by the λ-calculus and protected by the DNA structure of Capability Limited Addressing, can survive and evolve in hostile cyberspace. When digital media runs mathematically, software survives as an engineered functional machine. Each λ-calculus namespace is an independent machine, a software species privately in international cyberspace.

The λ-calculus builds the machine directly from individual functions, symbolically connected to create a self-standing private Namespace. Each Namespace can fearlessly co-exist. After a formal introduction, any two Namespace objects can interwork in cyberspace. The functional symbols are bound dynamically as threaded computations that calculate results secured function-by-function. Using Capability Limited Addressing these software objects are digital versions of the cogs and leavers in Babbage's two engines or the beads and rails of the Abacus. Exhaustively evaluated as atomic units, they remain reliable and survive as reproducible, fail-safe machinery for the entire life span of the underlying hardware. At the same time, as a programming language, they endure forever, crushing the cost of computer science by flawlessly protecting democratic society.

For example, consider Ada's mathematical program, which she created two hundred years ago. It is a λ-calculus function abstraction that could run today in the Mathematical Namespace of a Church-Turing Machine. None of the programs written by John von Neumann,

Butler Lampson, or anyone else could run on any of the latest branded computers. The short half-life of the branded software is one of the most expensive problems with General-Purpose Computer Science. It must be constantly patched, redesigned, and recompiled due to the arrival rate of outside corruption and incremental improvements, and it is overwhelmingly costly due to continuously retesting monolithic applications.

On the other hand, software in a Church-Turing Machine is like Ada's Bernoulli abstraction, the rails in an Abacus, or the scales on a Slide Rule. They are all gifts to the ages. They last. Their mathematics is flawless. Industrial Strength Computers offer this same flawless computational ability, allowing civilisations to coexist without debilitating international interference leading to war. Nations will prosper faster and further when isolated from terrorism and cybercrimes, Even more so if software as a function abstraction lasts generation after generation.

Consider the impact of industrial-strength digital software, supporting democracy instead of dictatorship worldwide. Digital guardrails, like the rails of an Abacus, will serve civilisation for thousands of years. It wins over all other alternatives because the software faithfully serves both functional science and the citizens of society. The science of the λ-calculus embedded in a Church- Turing Machine guarantees software survives every worst-case test of time.

The Slide Rule was a monumental gift to humanity. It was revealed not in 1614 by John Napier[59], the Scotsman who invented logarithms,

---

59   John Napier 1550 – 1617 was a Scottish landowner, mathematician and the 8th Laird of Merchiston. He invented natural logarithms to simplify the long multiplication he needed for astronomy. He also invented the decimal point and invented 'Napier's bones' as a tool to perform multiplication. His Latin publication, Mirifici Logarithmorum Canonis Descriptio (1614) the first logbook spread rapidly throughout Europe, contained ninety pages of hand calculated tables he spent twenty years compiling. Henry Briggs professor of Geometry at Gresham College, London visited Napier in 1615 and again in 1616 and they began to re-scaling logarithms to the power of 10. After Napier died Briggs continued the work and by 1624 his folio Arithmetica Logarithmica, contained thirty thousand numbers to fourteen decimal places.

but a decade later by William Oughtred. He replaced Napier's Bones[60] with the perfect function abstraction for multiplication, division, and every other mathematical function. It will remain so not just for millennia but for eternity.

Later, during the Industrial Revolution, Charles Babbage invented his two mechanical machines, the Difference Engine for polynomial functions and the Analytical Engine for programmed mathematics. Ada Lovelace explained how these mathematical machines could be programmed to compute and calculate as a software machine. She needed no operating system or any compiler. The machine language Babbage provided was mathematics, as written on a chalkboard.

Her program of twenty-five precise mathematical expressions calculates numbers from the Bernoulli series. Its self-standing function abstraction is a perfect software machine. As a mathematical function abstraction, it will last forever. None of these computers needs a branded operating system or any unfair privileges. There is no proprietary baggage to shorten the half- life of the software or mathematical voids for malware to cause trouble. Ada's program still works today, as it did so long ago. It is the gift of Industrial Strength Computer Science that software survives both logically and physically to the future of cyber society.

Consider Bronowski's Rose with replicated petals of natural beauty. It needs no third-party baggage favouring disease: centralized monopolies, criminals, spies, or enemies[61]. Ada's small self-standing function abstraction exemplifies a Rose in cyberspace. The λ-calculus in cyberspace matches the beauty of nature in the natural world.

---

60   Napier's bones form a manually operated calculating device created by John Napier for calculation of products and quotients of numbers. The method was not based in his logarithmic scale but on Arab mathematics and the lattice multiplication used by Matrakci Nasuh and Fibonacci's work.

61   Cyberspace is a WMD given to the worst instincts of humanity. Not just smash and grab criminals but a whole new criminal class that conspires against trusting democratic forms of society. The GRU in Russia, Mossad in Israel, experts at GCHQ, in MI6, or at NSA, CIA, FBI, etc. It only takes one bad apple with the necessary skills, and all is lost at once.

Ada used twenty-five punched cards to compute results, then (theoretically) printed by Babbage's unfinished Analytical Engine without error, human involvement, or centralized, third-party operating system baggage.

As a programmer, Ada used pure mathematics to define and run machine code on Babbage's platform. She composed just twenty-five scientific steps on paper or a chalkboard, and then she exchanged letters daily between herself and Babbage as two cooperating students sharing, reading, and understanding ideas; see Figure 5. No such possibility exists with the latest Generl- purpose computer. For humanity, mathematics using either a slide rule of the Analytical Engine is simple; it is fair to all, a level playing field for the endless future.

Cyberspace must be the same for the good of the nation and every following generation. As a λ-calculus computer, a Church-Turing Machine is universally fair because the laws of nature protect the future of software by flawlessly meeting the expectations of both science and society. A General-Purpose Computer is limited to Turing Commands that express only half of the Church-Turing Thesis. Consequently, hard-wired monolithic complexity is always one of a kind. Flexibility is centralized, enslaved to a unique compilation as a privileged virtual machine, only correct with the assumptions made at one moment.

Each centralized investment is short-lived and cannot be reused or survive until the next monthly upgrade. This software uses peculiar dysfunctional machine code, unlike λ-calculus, Babbage's Analytical Engine, the Slide Rule, and the Abacus. For software to survive, it must accurately reflect the statements of logic and mathematics written on a schoolroom chalkboard.

SYMBOLIC DIRECTIVES

A Church Instruction is symbolic and thus object-oriented, with every symbol on the chalkboard matched by an immutable Capability Key, accessing a digital object. The framework of context continuously redefines Turing's α-machine as a sensitive computational gene. For each frame, the form-matches function preserves the golden rule of natural selection and the universal model of computation that allows a rose to grow, birds to fly, and fish to swim and evolve generation after generation. Each computational frame is functional and reusable and takes nothing for granted.

In this generic architecture of dynamic life, symbols define a pervasive directional DNA hierarchy that keeps the dynamic computations on track. The functional mechanisms from the chalkboard are transparent in every detail, showing how every expression is composed. When implemented as a Church-Turing Machine, malware and hacking are exposed instead of hidden. Errors are immediately identified and removed in advance of any harm or catastrophe.

For example, a decimal number is the function abstraction of an abacus, which has, by design, beads standing for our own four fingers and a thumb. These beaded rails form an atomic machine that counts from zero to nine, where the thumb stands for five fingers. The framework of rails creates each function abstraction, an array of decimal digits to store significant, larger numbers. The Abacus is an *abstraction of abstractions*, where the form of the Abacus matches a large number that calls upon the rails as ordered digits.

Each rail is a functional symbol in the self-standing machine. Each one is a limited and protected unit for calculations—the beaded rails, as mathematical digits, only accept values from zero to nine. Thus, errors of scale are easily detected. The limited scope is easy to verify and confirm. The typed structure prevents atomic errors, and while human mistakes remain possible, a fail- safe thought process prevents unreasonable actions. Just as for students at school, all errors are recoverable. At the same time, outside interference is transparent and easily prevented by computational privacy and, when needed, locked doors— catching malware from another chalkboard or a fraudulent request on the spot.

IMPERATIVE COMMANDS

The procedural programs in a General-Purpose Computer use imperative commands that cannot detect mistakes or malware unless each test is hard-wired into the program in advance. The virtual machine is not even close to catching every error in a shared compiled image. The computations are opaque, bound together by the offline compiler and peculiar branded conventions. Furthermore, the assumption of blind trust first made by von Neumann undermines all the added baggage of centralized memory and shared operating systems. Static binding using imperative commands creates a compiled mathematical contraption where corruption is hidden, undetected, and unnoticed.

The compiler statically binds Turing instructions into imperative programs statically linked together as a monolithic virtual machine, but the software machinery using imperative commands is fragile and unreliable. It is not a valid machine but a bag of loose parts like a pile of stones. Like Napier's bones, they are hard to organize compared to the simplicity of the λ-calculus embedded into a slide rule. The sequential, static nature of procedures cannot detect a dynamic attack. Only dynamic binding, performed automatically by hardware in real-time, can prevent silly mistakes and deliberate, planned attacks. The λ- calculus, as a meta-machine, can verify and validate the mathematical symbols to detect and prevent corrupt actions on the spot.

The imperative Turing commands cannot create a self- standing machine. Sequential programs and imperative commands hide mathematical flaws. When von Neumann ignored the λ-calculus and overstretched Turing's α-machine, he deflated the raw power of symbolic (object-oriented) machine code, losing the natural dynamic powers of the λ-calculus to maximise computational performance on every front. He replaced the critical powers of scientific automation with bureaucratic, fallible human best practices. Best practices only grow evermore complex and untrusted due to inevitable human mistakes.

This flawed foundation even threatens the acclaimed 'block-chain' technology. In global cyberspace, one mistake in correctly and repeatedly applying best practices somewhere is a threat everywhere. It is an endless Marathon race with no winners, not even survivors. Imperative, statically bound procedures cannot scale or defend themselves from attacks, and every best practice assumption made by branded experts fails, one way or another, with the less-skilled public.

SYMBOLIC INSTRUCTIONS

A Church-Turing Machine is a symbolic computer using dynamically bound namespace limited instructions. A λ-calculus meta-machine encapsulates Turing's α-machine. The machine code is dynamically bound to symbols instead of linear page-based virtual memory. Capability Limited Addressing encapsulates each symbol in the λ-calculus with specified, functionally typed digital boundaries that enforce and sustain the media-tight laws of the λ-calculus as digital Variables, Abstractions, Functions, and Applications in a digitally coherent Namespace.

Each symbol is a λ-calculus variable. The encapsulated nuclear structures implement the λ-calculus types as mathematical expressions. The media is assembled into a digital form, encapsulating geometric boundaries, typically a base and limit memory address. When dynamically bound as object-oriented machine code, Turing's α-machine is object-type limited. The bound objects offer typed access rights for binary data, used by Turing's α-machine and immutable Capability Keys, used by the λ-calculus meta-machine as ordered by Church Instructions.

The defined mathematical expressions of a namespace match the symbols and preserve the Babylonian golden rule. Just as Ada invented programming as symbolic abstractions used since the beginning of time, expressions on her chalkboard define function abstractions in a λ-calculus Namespace:

$$Area = \pi r^2 = a$$

$$f(x) = a_0 + \sum_{n=1}^{\infty} (a_n \cos \frac{n\pi x}{L} + b_n \sin \frac{n\pi x}{L})$$

$$a = b + c$$

$$myConnection = Switch.\, Connect\, (me, myMother)$$

FIGURE 3: SCIENTIFIC EXPRESSIONS AS SYMBOLS IN A λ-CALCULUS NAMESPACE

These scientific, mathematical expressions in Figure 3 are like those Ada Lovelace proposed to program the Analytical Engine, how Babylonians used the Abacus, or how engineers like Isambard Kingdom Brunel designed his engineering feats using a Slide Rule.

The last expression above is from the PP250 telecommunication experience. Using such all-powerful statements, the object-oriented machine code of PP250 abstracted and simplified the global network to the single functional Church Instruction. This Church instruction needs four Capability Keys to Call the *'switch abstraction'* to use the *'Connect'* function and construct a voice call from *'me'* to *'myMother.'* In this context, *'me'* and *'myMother'* are function abstractions of two global telephone endpoints. Ada foresaw this beauty in the power of symbolic abstraction when building on Babbage's functionally flawless idea of computer science.

This application-oriented function is a high-level expression implemented in object-oriented machine code without needing all the compiled baggage of a branded General-Purpose Computer. Such a statement needs no superuser privileges, and if any existed, branded operating systems would block such a global statement. Furthermore, these limitlessly powerful expressions directly implemented by a Church-Turing Machine are also fail-safe.

Redundant hardware guarantees fail-safe software reliability as a fault-tolerant multiprocessor. For networked services, PP250 offered decades of software reliability and cybersecurity. The critical advantages are malware suppression, hacking detection, and preventing software break-ins and break-outs. The MTBF of a PP250 namespace exceeds the reliability of the networked hardware that protects against all single faults, both hard and soft.

When Ada Lovelace designed her algorithm for Bernoulli Numbers, her machine code inherited Babbage's *Infallible Automation*, which is inherent in the mechanical component of the Analytical Engine. The λ-calculus enforced by immutable capability tokens achieves the same component quality as digital abstractions.

A General-Purpose Computer only supports compiled procedures, hardwired as a monolithic component enslaved to the branded baggage of a particular supplier. All this software, including the operating system and the system administrators, is trusted blindly. Its monolithic process debilitates software quality in intractable ways.

- Functional modularity is nonexistent.

- Symbolic addressing is missing.

- Software boundaries and privileges are shared.

- Encapsulated atomic functionality is impossible.

- Machine code is complex.

- Compilers are needed

- Adding memory increases threats

- Recompilation requires a full regression test

- Blind trust replaces science

- Branded conventions prevent safe interworking

- Unfair privileges exist by default

- Peer-to-peer communication is blocked

- Errors and failures remain undetected

- Best Practices grow in complexity, and skills needed

- Propriety solutions inevitably conflict

- Hidden corruption cannot be avoided

- Excessive code is required to overcome static binding

- Monolithic software is fragile

- Evolution is slow and expensive

- Urgent patches destroy the invested value

OBJECT-ORIENTED MACHINE CODE

A Church-Turing Machine uses object-oriented machine code. Most importantly, the foundation is an object-capability system built into the hardware by a λ-calculus meta-machine. Capability Keys as DNA strings group the functional nodes in a dynamically bound implementation. The capability-based, object-oriented machine code corresponds to the symbols in a chalkboard expression—each symbol is a λ-calculus variable and an immutable Capability Key. Object access is grouped on a need- to-know, least-authority basis as defined by related scientific expressions. The symbols are deep-rooted as digitally protected implementations for each item as an atomic functional software abstraction.

This precisely defined mathematical assembly of object- oriented machine code organised as a directional DNA hierarchy is the formal algebra of a dynamic digital species. The nodes of digital media in the directional DNA hierarchy define the mathematical symbols of a scientific expression found on the virtual chalkboard that defines the Namespace.

For example, the symbols in Figure 3 have names and precise functional meanings. The list of symbols and the DNA of their relationships entirely describes the functionality of a namespace. Each is precisely defined and exhaustively testable as a λ-calculus functional variable and an independently assembled capability-limited digital object. The symbols translated by immutable Capability Keys unlock type limited access rights to an expressly limited and functionally typed computational space.

Pure data variables like *x, a, v,* and *n* offer Read-and/or-Write access for the encapsulated Turing Commands. However, the hardwired Turing commands are bound to symbols instead of pages in shared memory. This distinction makes all the difference and effectively costs nothing in time while removing all the baggage of General-Purpose Computer Science.

Algorithms like *-, +, *, /,* and *Switch* offer Execute access rights as needed for programs in any computer, including Turing's α- machine. Function abstractions are more complex. Examples like *myConnection, me,* and *myMother* are digital black boxes, frames of functionality and data, potentially linked to other function abstractions as a DNA chain. Function abstractions are executable capability keys protected by a semipermeable digital membrane called *Enter.* So {*Enter. someClass*} opens another C- list while saving the existing context on the capability-defined access rights thread stack.

The '=' symbol defines the Call instruction as a Church Instruction for a synchronous or asynchronous call (or a remote call) passing variables for another function abstraction to process.

A typed and limited Church Instruction unlocks the access rights to a symbol from the corresponding Capability Key. The Church Instructions dynamically and transparently bind the named media to a context register that constrains any Turing α- machine errors. The λ-calculus concept of an *application* is a Thread defined by the context registers listed later in Figure 12. Each symbol from the DNA string for the Thread includes a DNA string for a Function Abstraction. Capabilities exclusively define the frame, and the limited access scope of every computation defines the instance data used by the Thread and another for class and function defined by the current frame of abstraction.

For PP250, capability registers are the context registers to scope each authority, limiting the actions of the encapsulated binary computer to the simplicity of Turing's α-machine as a computational digital gene.

$$x, a, (v)^{n}, \sum_{v}^{w}, n, k, -, +, \sqrt{v},$$

$*, myConnection, Switch, Connect, me, myMother, =, \leq,$

FIGURE 4: A LIST OF CAPABILITY KEYS DEFINES THIS EXAMPLE NAMESPACE

The symbols from the chalkboard define each object in the namespace. Any remote namespace, device driver, and thread defined by Capability Keys can also reference local or remote function abstractions. The objects representing an active Namespace and the active Thread are also defined using reserved context registers shown for PP250 in Figure 12.

The symbols in Figure 4 represent a nonsensical namespace created by the expressions in Figure 3. A Capability Key identifies a slot in the Namespace table to locate the additional meta-data used by the λ-calculus meta-machine when unlocking access rights to the identified object. Each digital object is typed limited as binary data or a C-List of Capability Keys. A General-Purpose Computer only understands binary data. However, the five λ- calculus concepts of Variables, Abstractions, Functions, Applications, and a Namespace all exist in a Church-Turing Machine represented at run-time by capability context registers.

Binary data has no industrial strength to resist fraud and forgery and survive the tests of time when attacked by malware or a privileged human hacker, but Capability Keys are immutable digital objects. The immutable Capability Key is a trusted digital currency for distributed objects and networked information exchange that keeps keys as private secrets, preventing malware and hacking.

Safe access to data is guaranteed by the owner as defined by unique Capability Keys. Data is never stolen from the Namespace if a Capability Key is never shared with untrusted abstractions. The incremental assembly of each Namespace guarantees unknown objects are never included, locking malware out of each Namespace. The enforced boundaries of Capability Limited Addressing are digitally impervious and functionally restricted according to the needs of the λ-calculus.

These access rights are embedded in the need-to-know capability key to filter out all mistakes and prevent attacks. The media-tight configuration of function-tight and data-tight objects limits use to authorized users. Object by object and nuclear function by function, the limited access rights and the digital perimeter boundaries sustained by hardware offer in-depth and detailed protection, bringing law and order to the wild digital frontier of computer science.

Each Namespace is a structured assembly of modular digital objects. The Capability Keys define the directional DNA to match the expressions on the virtual chalkboard. A Capability Key can define a directly accessible atomic object, a complex structure of linked Capability Keys in chained C-Lists, of a remote Namespace.

There is no main program claiming the default authority of a super-user and all the superuser privileges. The DNA tree always limits the power to one functional node within one Namespace; no global authorities exist. Access rights do not exist without a key; thus, computational power and privilege start at zero. There is no default power to misuse. One namespace cannot access another without a formal introduction that implements the functional communication protocol for object-capability exchange.

The result protects the contours of digital media as an interworking algebra of mathematical symbols. The boundaries are skintight, minimize access rights, limited by a need-to-know, and limited to the smallest user group with appropriate case-by- case access rights. The namespace encapsulates a software machine defined exclusively by five λ-calculus concepts.

- Capability Keys are named λ-calculus Variables,
- A C-List structures function abstractions as object- oriented machine code in a DNA string,
- Functions are executable machine code objects,

- A Thread, as a λ-calculus Application, applies confidential data to a set of LIFO function abstractions,

- A namespace map of capability keys holds each key's algebra and geometry.

The multi-dimensional DNA framework with private Threads and public abstractions detects and prevents corruption while holding data private to a namespace. Finally, the algebra of Capability Limited computers adds lock-and-key security under the program and policy control of the namespace owner, converting the centralised dictatorship to a citizen's democracy.

CLOCKWORK CYBERSECURITY

Not only does Capability Limited Addressing protect software from itself, but most importantly, the directional DNA hierarchy extends a private security policy from the digital frontier into the hands of a namespace owner. In this context, Namespace is a capability-based desktop, including a virtual chalkboard. It acts as a secure browser[62] and a power box[63] for object-capability enthusiasts. These fraud and forgery-resistant structures use Capability Keys as immutable handles for each Namespace's subjects, objects, verbs, and variables. The security policy is exercised directly and indirectly through delegated Capability Keys and the directional DNA hierarchy of the namespace.

Capability Keys convert General-Purpose Computer Science's centralised, authoritarian nature into a democratic, edge- controlled solution. A capability-based computer deconstructs and distributes power to the data owner, placing the users in control. The user owns their confidential data while sharing (downloaded) networked services. Its edge control is vital for a citizen's cyber democracy in a 21st-century electronic village.

Binary data hides fraud, forgery, and corruption caused by digital interference. Binary data will never be private or secure in a centralised monolithic system controlled by service providers. PP250's experience shows that corruption by undetected software

---

62   The DARPA Browser investigation took place at Combex by Mark Miller and Marc Stiegler.

63   A PowerBox is a security pattern common in the Capability Security Model.

bugs in a monolithic compilation is immediately detected and purged in a Church-Turing Machine. The rapid detection of errors using Capability Limited Addressing accelerates the development cycle and significantly improves software quality. Updates are live as soon as the home location is updated. Dynamic binding avoids the recompilation process, avoiding a cumbersome delay and preserving the invested value in the bulk of the unaffected software. Finally, software changes take over automatically using the Capability Keys to maintain synchronization with the latest release.

A straightforward clockwork mechanism programmed by the handful of Church Instructions covers every worst-case complexity in cyberspace. For example, symbols can represent the access rights to remote operational namespace locations, device drivers, and parallel Threads. It extends the clockworks of the λ-calculus uniformly across cyberspace. All from one transparent, easily understood, thoroughly evaluated, scientifically validated, trusted, and certified clockwork foundation.

It meets the demands of the 21st century as seamless mathematics, symbol-to-symbol, right across cyberspace. Global applications can reliably span the world and remain mathematically exact, culturally sensitive, and true to form at every location. These computations are just as Ada Lovelace programmed the Analytical Engine. They are pure scientific expressions of functionality, from the decimal arithmetic of the Abacus to Ada's Bernoulli example, as shown in Figure 5, and beyond achieving her functional vision of computer-generated Art and Music.

In every case, the object-oriented machine code is both fail-safe and future-safe, with the Industrial Strength Computer Science demanded by a cyber society coping with a conflicted world disrupted by Artificially Intelligent Malware.

$$\frac{x}{e^x - 1} \equiv \sum_{n=0}^{\infty} \frac{B_n x^n}{n!}$$

FIGURE 5: THE BERNOULLI EXPRESSION EVALUATED BY ADA LOVELACE

A Church-Turing Machine is the universal, fail-safe, future- safe solution for cyber societies to prosper in the electronic village. Endlessly united by disputes and chained together by disagreements, immersed in the electronic village where cybersecurity is vital for survival. Democracy through edge- controlled privacy ruled by common law and order is the essential requirement.

The converged digital network of life-supporting services demands decades of reliability from a parallel processing global platform. It was still unrecognised and unaccepted by the per- electronic age when a flood of microprocessors for Personal Computing washed away the PP250 and all other competition, leaving only the broken results of General-Purpose Computer Science.

Everything from clickjacking to denial of service, from massive data breaches to Artificially Intelligent Malware breakout, can be solved with the Church-Turing Machine, including placing control in the hands of the citizens in democratic cyber societies. Putting the user in charge extends a people's democracy into cyber- space. A nation's law and order and other lesser groups must always be citizen-led using private namespace security policies.

In a conflicted world, digitally compressed as an electronic village, seamless digital functionality is vital to serving everyone fairly and equally; democracy demands this. The virtualization of time, distance, services, and society has uncharted consequences unless the citizens remain in charge of their destiny.

All prior assumptions and every custom or rule of society will change. Still, changes cannot include surrendering one's individuality or a nation's democracy due to the lack of machine integrity. While undetected crimes amplify threats and while criminals escape undetected, the unelected branded dictators will take control over cyber society, and it will not be a democracy. Already, rules that belong to citizens are changing, and new rules are emerging as best practices. For democracy to work and society to prosper, the assumption of General-Purpose Computer Science must end. *'Infallible automation,'* *'trusted software,'* and *'citizens rule'* are the buzzwords for 21st-century cyber society. Science and evidence tell us the universal model of computation driven by the λ-calculus is the way to achieve scientific machine integrity essential to safely Virtualise Democracy.

CHAPTER 5

The Ishtar Gate

The force that makes the winter grow
its feathered hexagons of snow,
and drives the bee to match at home
Their calculated honeycomb,
Is abacus and rose combined. An icy
sweetness fills my mind,
A sense that under thing and wing
Lies, taut yet living, coiled, the spring.
Jacob Bronowski (1908-1974)

Computer science began in the markets where the practical necessity to add and subtract large numbers started. The first machine to compute was the Abacus. It emerged from Southern Mesopotamia during the Early Bronze Age. As the power controlling trade along the Silk Road, Babylon was the centre of the ancient world; it became the cradle of civilisation, and this new technology, the Abacus, accelerated civilisation and prosperity.

Today, mud and brick remain the only clues to this first global village, erected fifty miles south of Baghdad, in Hillah. Once the Euphrates bisected this bustling centre of international commerce, the ancient town grew wealthy from trade in markets encapsulated and dynamically guarded by a locally enforced Babylonian security policy. When the river shifted, the western side flooded, and Babylon returned to dust.

This birth of Civilisation included an everlasting gift to science and society. It was the Abacus and the Babylonian Rule of computer science that physical form matches mathematical function.

The Abacus as a machine supports two mathematical functions in a shared form. The form and function unite as the Babylonian Golden Rule. Chained abstractions add a directional DNA hierarchy, first for a single decimal digit. When replicated as an array, a sizeable decimal number exists. Each decimal digit is stored on a single rail to abstract counting like the fingers on a human hand. Children worldwide learn to count this way. The physical array abstracts a decimal number exactly as written on a chalkboard. The Abacus is a mathematical extension of mechanical hands for the easy addition or subtraction of two large decimal numbers.

The directional DNA hierarchy, as seen in Figure 6, is open- ended. Scientifically, the size 'n' can grow to one less than infinity. As an atomic abstraction of a human hand, it frames a scalable machine of ordered decimal digits representing any significant number. The design serves two masters. It faithfully serves society by flawlessly simplifying and serving the counting functions of add and subtract. Millennia, after millennia, perfect results guarantee continued prosperity and everlasting survival.

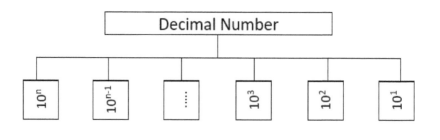

FIGURE 6: THE MATHEMATICAL DNA OF DECIMAL ABSTRACTIONS

Like Jacob Bronowski's Rose, the machinery of the Abacus hides a coiled spring of Nature within each abstraction. They are both computational genes that navigate their mathematical DNA to achieve a result, but only a Rose is infallibly automated. Human error contaminates the consequence of the Abacus. Nevertheless, without exaggeration, the Abacus remains an everlasting wonder, a monumental achievement of the Ancient World. It remains an eternal gift to every later generation and every civilisation. Digital computers must be the same flawless gift to the future of humanity, society and democratic civilization.

TRUE-TO-FORM

Both the Abacus and the Rose are, in their ways, true-to-form implementations of mathematical science. Nothing beats the beauty of a Rose or the simplicity of the specialised and precise decimal arithmetic of the Abacus. Not duodecimal arithmetic, not binary arithmetic, and certainly not general-purpose arithmetic, if indeed there is such a thing. The Abacus specialises in the flawless perfection of decimal arithmetic. The specialised laws and order of decimal arithmetic require a stubborn, specialised framework. The beaded rails are the replicated petals of an Abacus. Each rail has a substituted property as a digit. For a computation, a second digit leads to a new result. It is an atomic media machine, just as a Rose is another specialised form of life with another directional DNA hierarchy.

Like every living system, the Abacus matured through natural selection and stone piles were discarded. As the best solution, it evolved faster for ever-larger numbers, prevented interference, and survived the tests of time. Survival and rapid evolution are the fundamental capabilities of a species that thrives. Engineered by trial and error, the Abacus outlasted all competition until the pocket calculators of the electronic age. A clockwork algorithm scales any number of rails to model decimal arithmetic correctly. Natural selection simplified and extracted the best implementation. The result is a scalable machine. A machine that uniformly grows to any size of a decimal number. Powered by the universal model of computation, substituted private variables apply to each mathematical or functional abstraction.

Each rail is an atomic copy of a true-to-form mathematical abstraction, creating a Namespace with the DNA for a significant decimal digit. Like the petals of a Rose, each abstraction is a generic atomic 'gene-machine' with separately substituted variables as private properties that define the size and the colour. The primitive form is perfect for survival as an arithmetic computer; nothing is superfluous. No unnecessary baggage exists since the Abacus is not general-purpose but dedicated to perfect decimal arithmetic. However, the mechanisms follow the universal computation method of science and nature. The mechanics abstract a human hand, where a clenched fist is zero, and an open hand represents nine, where the thumb is a mathematical symbol, a type that represents five fingers.

Like the hand, the rail of an abacus cell counts from zero to nine, as is a size-limited machine. It constrains the functions to one decimal digit. Like a Rose petal, the structure obeys a hidden DNA string that defines the blueprint as the mechanics of each function in an abstraction. The DNA string is replicable as reproducible media. Every living thing is this same way, built from a single atomic frame of a nuclear machine called a gene in nature, programmed by the clockworks of the DNA chains. The hidden clockworks are a coiled spring of chained programs within the framework that operates through the pure force of Nature.

When logarithms were discovered in 1614 by John Napier, the Slide Rule followed, once again, through natural selection. The speed of evolution, this time as scaling functions, powered the Slide Rule to quickly and permanently replace Napier's Bones[64]. The form of the Slide Rule also changed as a specialised machine for scientific scales. It works alongside the Abacus for a different species of mathematics, making two significant improvements towards Babbage's hallmark of *Infallible Automation*.

---

64   Napier's Bones is a manually operated device created by John Napier for calculation of products and quotients of numbers. The laborious method is based on Arab mathematics and lattice multiplication. Samuel Pepys acquired a set of Napier's Bones as he recorded in his diary entry of Thursday 26 September 1667.

- This time, scientific scales guarantee each algorithm, avoiding human errors and extending the supported functions. Trained skills are human hand and eye coordination to substitute the variables correctly from the scales to reach a scholarly result.

- Second, embedded algorithms are Bronowski's hidden springs built into one generic machine. A Slide Rule is *'multi-functional'*; in that sense, it is a generic class. It is a general-purpose, object-oriented class of function computation.

Again, the public learned to use the Slide Rule for the remaining core of essential mathematical functions. Both machines use the Babylonian Golden Rule of form matching function. This rule flawlessly performs mathematics through uncompromised computer science found in nature's universal model of computation that governs every living species.

When applied to computers, the golden rule shapes the dynamic limits of mathematical calculation through the universal computational model driven by the λ-calculus. The science of λ- calculus brings specific laws to structure and bind an ordered calculation. In a Church-Turing Machine, Turing's single tape α-machine is a computational gene for a mathematical DNA of function abstractions. The symbols in λ-calculus are the petals of mathematics. The DNA relationship between the symbols is the rails of a fail-safe, reproducible computation, another species of mathematically calculated dynamic life.

A directional DNA hierarchy of symbols in expressions on a chalkboard guides the calculations to an exact result. The Church-Turing Machine encapsulates these calculations as computational frames picked from the available DNA string.

Each frame is a mathematical abstraction afforded the same tangible dignity and mechanical integrity as an Abacus's solid wood, rails, and beads. Using Capability Limited Addressing the functional simplicity and ease-of-use with the same long-term stability, survivability, and speed of evolution is immediately translated into the digital world as object-oriented machine code—lightweight fail-safe instructions without any general- purpose baggage.

The goal is *Infallible Automation*. A safe and secure software machine that connects and communicates flawlessly across Virtual reality. The software machine is as beautiful as a rose, flawless as an abacus, multi-functional, and easy to use as a slide rule. In a Church-Turing Machine, the software faithfully serves humanity and science. Like the Abacus, these mathematical software machines, emphasising software, could last for thousands of years. Using the Babylonian Golden Rule, even super-intelligent (AI) software is kept in line with the level playing field.

SURVIVAL

As the product of humanity, a converged digital society is unavoidably hostile, expressed by malware, hackers, and soon malicious Artificially Intelligent Malware. Combatting hostility requires industrial strength. Computer science must apply the Babylonian Golden Rule using capability-limited addressing to reproduce Babbage's hallmark of *Infallible Automation*.

Capability Limited Addressing detects and prevents the human-inspired hostility of malware and hacking. Only then will the software be everlastingly fail-safe. When software survives, like Ada's program, for generations on end, computer science will be an everlasting monument to the progress of humanity and the democratic future of the 21st century. The survival of all virtualised societies as cyber-dependent nations depends on this level playing field where citizens democratically control their digital shadows and express their private cyber destiny.

Conflicted cyber societies can only prosper and survive side-by-side when malware and hacking threats disappear. Progress in the cyber age must be through fair competition, as a level playing field of mathematical laws and order. In Babylon, the first step took place, using mathematics as an everlasting machine language. Simultaneously, it served science and society. A clockwork form of arithmetic simplified and perfected to the specialised essentials to flawlessly enforce nature's raw powers.

The purity of an atomic machine obeys the Babylonian Golden Rule as the flawless model later defined by Alonzo Church as the λ-calculus. The endless beauty of replicated, recursive, functional mathematics implements the Church-Turing Thesis as a Church-Turing Machine by applying and obeying the laws of λ-calculus.

ENCAPSULATION

C ivilisation prospered through the Abacus. It all began in a peaceful city that supported art and science—achieving encapsulated security through trusted physical boundaries with locked and guarded entry gates. Encapsulated security nurtured the progress of civilisation. Babylon prospered because the framework of the Abacus enforced the law and order of arithmetic. Crimes were limited by checking visitors on arrival at the entrance gate to Babylon. From the start, science advanced, art was stunning, and enemies were overwhelmed. To the North, the fabulous, blue-tiled entrance gate, known as the Ishtar Gate, provided dynamically guarded access rights to the City and a beautiful Processional Way that protected and secured the walled city. Armed guards acting for the enclosed society check arrivals proactively for taxes, permissions, and the authority to proceed. This level of protection for the machine instructions in a General-Purpose Computer is missing.

The Ishtar Gate is named after the goddess of love and war and is decorated by the most brilliant blue and yellow tiles that artistically depict bulls and dragons. These were the symbols of the city's god, Marduk. The entrance gate proves, beyond doubt, the wealth, artistic talent, technological competence, and the remarkable cultural strength of Babylonian society. As a Wonder of the Ancient World, it remains as impressive.

The beauty of the Ishtar Gate, pictured on the cover with a beautiful Processional Way, is a magnificent monument to the birth of civilisation. As they once stood in Babylon, they are an impressive combination of engineering, art, science, industry, culture, security, authority, and power; the Ishtar Gate strikes dread and admiration in all who pass through this guarded entrance to the birth of peaceful civilisation. They now stand in Berlin at the Pergamon Museum. Beyond all doubt, the everlasting beauty and the detailed artistry certify the engineering talent of artists, who were also skilled scientists and leading-edge technologists.

They are engineered monuments, including the Abacus, to the skill and vision of an age who built things to last. The enlightened population prospered commercially, scientifically, artistically, and peacefully within the functional encapsulation of boundary walls and dynamically guarded access rights guaranteed by the Ishtar Gate. The enclosed boundary promised peace and security for those who lived under Babylonian law and order. The city empowered businesses to thrive, while trade and commerce flourished through the Abacus. The town prospered as the markets grew, and trade between civilisations soon extended Worldwide.

Babylon was the first global village, an international city of families, friends, neighbours, competitors, visitors, foreigners, spies, thieves, valuable goods, cottage industries, and dynamic commercial trade. It was where all coexisted and prospered side- by-side because the law, order, and the Abacus could all be trusted. Trust is essential to the continued prosperity and survival of civilised society.

MATHEMATICAL MACHINES

Babylon discovered that business grows by preventing crime. Merchants from the West quickly and safely bought silk and spices from the East. They traded and bartered all day long as a multicultural society. The languages were diverse; the translations were a problem. But like music, arithmetic is universal, and decimals begin with everyone's four fingers and a thumb.

However, one hand is limited, so counting continues to ninety-nine with two hands. Thus, the Abacus evolved to abstract even more significant numbers as a machine of many hands. No number is too big for a larger Abacus. The result over piles of stones was more straightforward to trust because it was transparent and understood. Thus, trades were quickly and confidently closed. Mathematically speaking:

1 finger = 1

1 thumb = 5

1 hand = 4 fingers + the thumb = 9

Unlike branded binary computers, instead, like music, the laws of mathematics transcend culture and spoken languages. Nature defines a universal model of computation that permeates life and science. The atomic machinery of the Abacus implements two of the four essential functions of mathematics automated by Babbage's Analytical Engine.

$$function\ (a + b) = c$$

$$function\ (a - b) = d$$

$$function\ (a/b) = e$$

$$function\ (a * b) = f$$

The Abacus and the Slide Rule demonstrate the primary mathematical functions are limited but highly adept. The standard resident functions[65] are enough to design a rocket to the moon and back, as well as the Boeing 747. The German Wernher von Brune used his Nestler slide rules bought before WWII throughout the NASA Moon Race.

The four core mathematics functions satisfied Babbage's numerical needs and Lovelace's symbolic vision—an inspiration going well beyond numbers[66].

As Menabrea said at the time:

> '.. limit ourselves to admitting that the first four operations of arithmetic, that is addition, subtraction, multiplication and division, can be performed in a direct manner through the intervention of the machine. It granted, the machine is thence capable of performing every species of numerical calculation, for all such calculations ultimately resolve themselves into the four operations we have just named.'

In this way, Ada Lovelace programmed her 'machine code' mathematically because Charles Babbage fully implemented these functions flawlessly to at least fourteen decimal places. Ada documented her pioneering Bernoulli program in Note G, the last of seven invaluable detailed explanations she added to her

---

65   See https://www.math.utah.edu/~pa/sliderules/ for a list. The Faber Castell Novo Biplex 2/83 N has 30 scales and 11 cursor marks.

66   It was while contemplating the vast field which yet remained to be traversed, that Mr. Babbage, renouncing his original essays, conceived the plan of another system of mechanism whose operations should themselves possess all the generality of algebraical notation, and which, on this account, he denominates the Analytical Engine - L. F. MENABREA

translation of Menabrea's earlier report in French[67]. It took only one page of chalkboard statements aligned with the machine settings and conditions. Ada needed no abstract language beyond simple mathematical statements, no central operating system, no superuser, and no unchecked source of human error.

Ada's symbols are mathematical functions, not just numeric values used by binary computers. These λ-calculus variables are far more potent than numeric values. Her algorithm is programmed functionally and is far more advanced in powerful statements than a general-purpose computer.

The symbols atomically separate the data from the functions. The application functions and dynamic data belong to computational Threads in the Namespace, while generic functions are available from service providers.

The λ-calculus variables are all mathematical types. The operator must memorize the function abstractions for addition and subtraction in an Abacus, using them repetitively for different values on each rail. At the atomic level, the calculation protocol follows the λ-calculus. It is indeed general-purpose, but a framework exists to implement the DNA, and the memorized algorithms are special-purpose, precise, and specific in every detail just for simple arithmetic.

Replicating the function abstraction for additional decimal digits is as natural as replicating petals in a Rose. The replicated rails share the same two memorised algorithms and compute numbers far more significant than nine in a cycle. Replication is simple but essential to both scalability and survival. Easy replication is part of an evolutionary life cycle driven by a directional DNA hierarchy. DNA is the blueprint to simplify reproducing complex life and dynamic calculations in hostile environments.

---

67   Ada translation including her Notes A-G and her first program is found here. Sketch of The Analytical Engine (fourmilab.ch)

Simplicity is important. For example, a decimal number is only valid between zero and nine, but adding or subtracting can result in an overflow or underflow signal, as learned at school. The event is carried to the next digit in the array by a computational Thread that cycles over each digit.

In a Church-Turing Machine, an immutable Capability Key represents each λ-calculus symbol in an expression. Each variable from the mathematical expression has an access right. Binary data allow *{Read.binaryData}* or *{Write.binaryData}*. The Capability Keys with type-specific access rights detect and generate errors on contact. On-the-spot error detection speeds up the development of the highest-quality software by detecting mistakes from the first experience. The typed access rights for each PP250 object are listed later. See Figure 16.

For example, an algorithm is a coded procedure defined by an *{execute.aNamedProceedure}* capability key, while a class is a C- List defined by an *{Enter.aNamedClass}* capability key. Each Capability Key hides the object details as a black box of type structured digital information. Each class has a private C-List storing the DNA string for one node in the namespace. Each node is a protected functional class, an atomic digital machine. It is activated by an *{Enter.theNamedClass}* Capability Key that acts as the dynamically guarded Ishtar Gate of the class. A call to unlock the gate is a guarded call that unfolds the coiled springs of executable machine code hidden by the type of limited Capability Key.

Each Capability Key is a token representing a mathematical symbol that includes customised access rights and the object's identity. The Capability Keys enforce the rule of *'need to know,'* and the access right applies the law of *'least privilege.'* These two rules are vital to any security system.

Spying was a significant concern during the Cold War as a centralized solution, Identity Based Access Control, developed in the pre-electronic age of individuals and physical organizations. So, General-Purpose Computer used individual names, private passwords, and Access Control Lists. But now, in the electronic age of digital convergence, individuals do not represent the level of detail in network attacks downloaded on demand. Networked attacks bypass identity-based security monitors and access lists. These attacks always succeed.

The Church-Turing Thesis addresses all networked back doors and open window problems because security works at an atomic level. Malware cannot bypass this nuclear security system. Every atomic object is capability-limited. Every download is constrained to Namespace rules without access rights when joining a capability-limited Namespace. The λ-calculus defines the networked foundation for *Infallible Automation* right across cyberspace, and the Thread mechanism defines a Machine to enforce cybersecurity that cannot be sidestepped or bypassed.

The Capability Keys deconstruct identity-based security and virtual machines inherited from the Cold War into atomic security, where the symbols of λ-calculus simultaneously define modularity and cybersecurity. In the digital world, Capability Keys are indispensable, not a luxury. They define access rights to enter a gate like the Ishtar Gate. Gates, like doors, are easily locked and guarded. Together with walls, they protect a safe place to store valuable and confidential objects. The rule applies to the distributed domains in a Namespace and cyberspace. Access rights and procedural guards must approve all visitors.

A THREAD OF A COMPUTATION

When using an Abacus to add or subtract, a human recalls the programmed algorithm and applies their variables confidentially. The framework maintains the security and privacy of the data. Its exact form of individual privacy applies to a computational Thread.

All around the world, Threads work in parallel, each with a private (wooden or digital) data structure to reach a private result with locally substituted variables and an identical generic mathematical function. The algorithms vary according to the abstract class. The DNA framework simplifies reaching the correct result because the immediate detection of errors avoids the defensive coding effort. Some Threads might perform tasks immediately as synchronous actions. Others might use an asynchronous solution that calculates partial results in parallel. The choice is one of performance, but the result is the same.

Mechanizing computational thread towards *Infallible Automation* came in stages. At first, humans remembered everything with the Abacus. The Abacus as a machine was used for convenience in structuring frames of large numbers, avoiding silly mistakes, remembering partial results, and supporting parallel activities. The algorithm unfolds the frames for calculation, one after another in sequence or by 'forking' the initial Thread into parallel activities that return partial results. The frames of calculations take control of the Thread until finished by returning a result. The hardware frames act as guard rails to verify, validate, dynamically constrain and remember valid results.

The machinery changed with the Slide Rule because the computational Thread stores the algorithm. Still, the thread remains a private calculation retained by the machine designed to simplify the computation.

It was circa 1622 when William Oughtred discovered the Slide Rule. He aligned two logarithmic scales, embedded as the algorithm for multiplication and division onto the two sliding shafts, and added a sliding cursor to find a result. It created another specialised type of flawless mathematical machine. The mathematical type is a λ-calculus function abstraction that encapsulates the mathematical algorithm embedded into the machine.

The λ-calculus guarantees the correctness of the algorithm that no longer needs to be learned by the human user. Indeed, alternative scales are embedded side by side to add other functions with equal simplicity, making the Slide Rule an object- oriented 'Class.' Learning to slide the rods is all the skill needed since the class algorithms are already correctly and reliably in place.

This generic form of a Slide Rule is an object-oriented class of mathematical computation using one of the embedded scales as the algorithm, but it still requires human operation. Still, progress towards *Infallible Automation* made a huge step forward. The Industrial Revolution and the Moon race owe their success to the mathematical magic of the Slide Rule. The Slide Rule abstracts the functions and values, allowing their substitution as embedded and encapsulated λ-calculus variables. The mechanical construction of function abstraction as an object- oriented class of machinery was a monumental step forward.

In Circa 1830, Charles Babbage reused the same object- oriented, form-matching function, Golden Rule, in his unfinished mechanical computers, the Difference Engine for polynomials, and later the Analytical Engine, the first fully programmable computer.

The object-oriented computations of the Abacus, the Slide Rule, Babbage's two engines and Turing's α-machine as a single tape machine are all cut from the λ-calculus cloth. For example, Ada's 25 program steps define an object-oriented class as an abstraction for Bernoulli Numbers. A chosen function abstraction returns one number in the Bernoulli sequence in response to a request for the item 'n.' Her Bernoulli class is a computational frame of mechanics implemented in mathematical programmable statements, creating a λ-calculus function abstraction.

In a Church-Turing machine, Capability Keys define each algorithmic frame {Execute.BernoulliFunction(n)}

Ada needed twenty-five mathematical steps as a sealed, object-oriented class of mathematical expressions.

Eventually, in 1936, the theory of λ-calculus clarified the practice, and the science formalised the two sides of the universal model of computation. The Church-Turing Thesis combines the work of Alonzo Church with his student Alan Turing, using the λ-calculus as a digital version of the universal model of computation first unearthed in Babylon.

As with every earlier achievement, a computational Thread controls the universal model of computation. The private thread drives the object-oriented actions of the chosen functions published by the class, providing the private values needed to calculate a result that returns to the thread. The stepwise actions that call on a function that returns a result could now be automated.

In Turing's α-machine, a paper tape provided the algorithm; the neat thing is that the machine is always the same. It is a General-Purpose Computer. However, the disadvantage is the visible framework that keeps each algorithm and the step of the algorithms on physical rails evaporated. The General-Purpose Computer allows accidental, unnoticed errors because the open- ended functional framework of the λ-calculus is missing— replaced by statically bound numeric variables.

DYNAMIC BINDING

By dynamically binding digital information to λ-calculus symbols instead of shared linear memory, the missing framework, the same computational framework that simplified computers, from the Abacus and the Slide Rule to the Analytical Engine, is restored. Dynamic binding redefines linear static memory into the mathematically woven fabric of the λ-calculus.

The same universal model of computation is used throughout nature as life cycles for every living thing. The dynamic protection mechanism of λ-calculus variables encapsulated the Turing commands in private parallel Threads to the chain of function abstractions selected by the universal computation model from each species' DNA.

Now, the simple, parallel replication and recursive cycle from the Abacus with the dynamic functions of an object-oriented class like a slide rule applies to the object-oriented machine code of a Church-Turing Machine. Even more significant for the information age is that the user always controls the access rights to confidential data. They can prevent data sharing—the opposite of the general-purpose computer that exposes everything as shared memory.

As first found in the Abacus, parallel processing is a generic advantage of dynamically binding the symbols as object-oriented machine code. Parallel threads in a Church-Turing Machine share the

Namespace class hierarchy of functional algorithms by substituting private variables in each thread. Threads work together and can be synchronised, sending messages using Dijkstra flags as a shared function abstraction.

None of the complex and opaque baggage invented by General-Purpose Computer Science is required. No shared virtual memory, offline compilation, proprietary binary conventions, monolithic binary images, unfairly privileged operating systems, complex best practices, or hackable superuser accounts are needed.

While a pile of stones could implement a general-purpose computer for duodecimal numbers or binary numbers, the procedures needed are irrational, complex to learn, and obscure because they add no value to the intended task. Without a computational framework, silly mistakes happen far too quickly in the opaque implementation. Fake and flawed results threaten the stability of the marketplace because facts are questioned when errors are undetected, and *Infallible Automation* is missing.

Trust falls as a direct consequence of malware and hacking that thrives. Because functional boundaries are missing, markets degenerate into escalating, unpreventable human conflicts. Unavoidably, given exploitable access rights, humans will corrupt a General-Purpose Computer.

In the Babylonian marketplace, a general-purpose pile of stones could only lead to fistfights. In a General-Purpose Computer, these same human errors lead to undetected crimes, international conflicts, loss of trust in government, expanding cyberwar, and contaminated False News. Human weakness undermines the fight to survive. In the mainframe age of the Cold War, however attractive a General-Purpose Computer seems, it hid schizophrenia that causes trouble and leads to conflict in the age of digital convergence. Opaque uncertainty that corrupts the progress of civilisation.

A core problem is unreliable binary data. It is easy to change, frequently misunderstood and leads to undetected corruption. The λ-calculus framework prevents this unavoidable weakness when using an overstretched Turing machine as a General- Purpose Computer. Binary data cannot be reliable without the digital security of λ-calculus enforced by capability boundaries. For example, a file path on the internet is unprotected as a series of characters in a string. These internet links can be easily corrupted, changed, or forged, as the file path triggering Norman Hardy's *'Confused Deputy Attack.'*

Misuse of binary data leads to every form of attack in cyberspace. The wide use of the *'Confused Deputy'* is the malicious exploit of choice. The attack penetrated Hillary Clinton's 2016 presidential run and shut down Sony Pictures in 2014. Indeed, there are so many attacks they cannot be counted. Junk emails received each day are high-risk, leading to increased text messaging. However, Texting does not solve or hide the problem of sending a link to a contaminated server. Because of this hidden threat, delete junk email and text messages before opening them. It makes the whole purpose of computer science and digital convergence a threat to progress.

Any software service that connects two machines through unscientific, branded conventions is at risk. A binary request cannot be validated or verified, and fraudulent requests from unknown endpoints are an attack vector that a patch cannot plug because the basement technology of a binary computer is dangerously flawed. Hackers and criminals make crafted requests using fraudulent binary variables in infinite ways. For example, a forged file pathname, specially crafted image tags, hidden web forms, hidden JavaScript, and disguised XMLHttpRequests are systemic digital engineering problems in General-Purpose Computer Science.

These software interworking requests go ahead without human interaction or infallible error checking. Using a browser is always dangerous. The cross-site scripting attacks used by criminals, Russians, and the North Koreans exploit trusting users, and the results are always tragic. Clickjacking attacks now dominate the corruption of innocent users. It all starts with corrupt binary data and a fractured mathematical model of computation that cannot scale.

A Thread as a universal model of computation with object- oriented machine code and Capability Keys to Capability Limited Addressing solves these problems. Hidden data cannot be misused. The atomic software machinery of a Church-Turing Machine adds the immutable Capability Keys to enforce reliable communication and prevent the malware attack that pervades dictatorial General-Purpose Computer Science and directly harms innocent users. Citizens must understand this threat and demand a solution. A national debate is urgently needed.

THE DNA HIERARCHY

The DNA of a species encodes the interworking relationships between the atomic cells and functional organs of an application. Using the symbols in the λ-calculus and Capability Limited Addressing the interworking object-oriented machine code is engineered and made fail-safe by a Church-Turing Machine. The object-oriented machine code is structured, calibrated, qualified, and trusted like the mechanical components in a clock.

Software needs the guarded modularity of capability- structured object-oriented machine code because it requires hardened, reliable components like any reliable industrial strength machine. The malleable nature of compiled binary images has unacceptably low MTBF. It is unreliable. It is unacceptable as the dependable substance of global cyberspace to automate civilizations' eternal future.

Proprietary conventions and binary data limitations allow undetected fraud and forgery to pervade cyberspace. It leads to a loss of trust and disagreements on every subject. As Marshall McLuhan pointed out, the *medium is the message*. If the medium is corrupt, the message cannot be trusted. Avoiding mistrust and disagreement is vital to civilised progress.

With mathematics, there are no exceptions to scientific truth. Only a Church-Turing Machine guarantees the mathematical results are always fail-safe. Error detection is vital for *Infallible Automation*. The architecture for Industrial Strength Computers as a Church-Turing Machine began with the PP250 but disappeared in the evolution of the microprocessor.

Trust in civilisation will only grow through the quality of shared engineered accomplishments. General-purpose computers only qualified as a scientific accomplishment as a stand-alone, batch-processing mainframe. The branded compromises and proprietary baggage of flawed assumptions spoil the science in a hostile network. The digital framework of a Church-Turing Machine using Capability Limited Addressing fills the void and restores the science of the λ-calculus that von Neumann ignored.

Using science will not kill industrial competition. Functional performance, size, and cost details differentiate a competitive range of Church-Turing Machines. Competition enhanced the Abacus and the Slide Rule; the same applies to Church-Turing Machines. The Abacus and the Slide Rule framed practical science with function-specific variations. A Church-Turing Machine is also open to forms of competition that reliably interwork using alternative capability forms that PP250 called *{Passive Capabilities}*. These capabilities are immutable values, not just constant numbers but also Namespace-defined capabilities and industrial-standards groups, not to mention programming language groups.

The functional architecture of a software machine is a chain of dynamic atomic calculations framed as a directional DNA hierarchy of Capability Keys gripping digital objects as a hierarchy of individually framed calculations. The frameworks of these function abstractions are the universal computation method of the private computational gene of a Church-Turing machine. The gene uses the universal model of computation to compute private objects from real-world events as a set of λ-calculus variables framed as a digital Thread. The private event variables represent the oxygen flowing in a digital bloodstream, framing an instance of the species as a machine. The directional DNA hierarchy of Capability Keys holds everything in place, subject to the context controls dynamically enforced by the λ-calculus using the context registers of a Church-Turing Machine.

A Turing-Command has no private rails to follow. They all share a common linear address space. They are bound to compiled pages in virtual memory, not the underlying mathematical models. The mathematically structured, directional DNA hierarchy targets the Turing-Commands to individual symbols in a mathematically scientific model. Its digital framework holds each calculation true to form by following the laws of λ-calculus and the directional DNA hierarchy.

To fight against cybercrime and the indiscriminate use of confused deputy attacks, monopoly suppliers are building digital fiefdoms as proprietary clouds. Software capabilities support a specialised operating system model.

It remains clear that one mistake by a privileged user, one late upgrade, one undetected hacker, one successful malware attack, or a missed best practice still has severe consequences. Not just because of the initial attack but because there is long-term harm with endless repercussions. Each undetected bulk data attack echoes repeatedly and surges forward every time a stolen credential is used in a subsequent attack.

Only when discovered, long after the first attack, is the software patched, but the material harm already caused remains unresolved. For a virtualised society living in an electronic village, this is lethal. Sooner rather than later, democratic law and order evaporate, replaced by branded dictatorships that steadily increase their authoritarian control over their branded, fragmented section of the marketplace.

Too many attacks remain unrecognised, destroying the foundation stones of government and society. The harm caused by accident, misunderstanding, or intent hardly matters. The result is the same. Both the medium and the message of General- Purpose Computer Science are unacceptable. Dependable, trusted software is essential for 21st-century cyber society to work if trustworthy cyber-democracy is to be achieved.

TYPED MACHINES

Over 4,000 years ago, through trial and error, the universal model of computation began with the trusted Abacus. The Abacus is not just the first computer; it demonstrates a perfect machine for private computation of decimal arithmetic for unskilled citizens living in a civilized society.

The machine applies the same universal model of computation used by every living thing in nature. Pure mathematical computation (à la the Abacus) removes disputes and unwanted baggage, protecting confidential and binary data from misuse and abuse. It is a trusted, encapsulated modular process that locks out malware, hacking, and unfair privileges while detecting program bugs at once on the first appearance. At the same time, the power of typed numbers becomes rails to keep dynamic calculations on track at extremely high speed with the added power of the λ-calculus, an anonymous function, with LISP or Haskell and other functional languages implemented in high-performance, object-oriented machine code.

Typed frameworks enforce mathematics as structured arrays of obedient function abstractions. The recursive, replicated, cascaded concepts simplify and scale the processing to any size or complexity with parallel threads of object-oriented machine code chained together as abstractions of abstractions. The benefit is a straightforward alternative model of computation. In addition to a Turing Machine assembly language, the Church Machine adds a λ-calculus high-level language driven by names in the directional DNA hierarchy.

The mathematical types of object-oriented machine code guarantee that the pure scientific structure will survive, like the Abacus, forever. A Slide Rule confirms this fact of nature exploited by the universal model of computation. The logarithmic algorithm never changes, and the matching machinery means the program can survive, dynamically bound to the instance variables for each calculation. Once again, the Babylonian Golden Rule proves essential. As faithful servants of humanity and science, computers must use as object-oriented machines to be future-safe.

The λ-calculus constrains the smartest Artificially intelligent malware. Malware, like terrorists, is locked out of the cockpit. No default privileges exist, and each object is trusted. In a Church- Turing Machine, hacking, malware, and unfair privileges vanish. The software runs on scientific, dynamically constructed rails. Capability Keys distribute authority with functionality according to a plan, controlled by the user of a Namespace as a citizen of a cyber civilization.

THE CLOCKWORK META-MACHINE

Each of the monuments to civilisation survives the tests of time. Like the Great Pyramid of Giza, monuments survive. The best ones are eternal. They are function-safe and data-tight. Technologically organized to pass every test of time, tailor-made as an object for an impervious functional purpose. As a computer, starting as the Abacus, each case has an encapsulated computational form to match a chosen computational function. In capability-limited cyberspace, protection is both physical and logical. Type-limited boundary checks guarantee the form of the object throughout time, and type-limited capability keys guarantee approved access rights. Thus, however small or large, every encapsulated object is isolated from malware dynamically guarded with the exact mechanisms used at the Ishtar Gate that once protected Babylon as civilization began.

At the beginning of the information age, Turing's α-machine as a binary computer worked in the locked rooms of the mainframe age, supported non-stop by trusted IT staff. Beyond batch processing, the software of the 21st century took off, starting with the PC and powerful microprocessors, and when the locked room and the dedicated staff disappeared, hacking and malware flourished.

Open-ended cyberspace emerged, as researched in 1936 by Church and Turing. Their Church-Turing Thesis builds on the λ-calculus, the symbolic science of open-ended cyberspace. The λ-calculus, as nature's universal model of computation, is continuous. Substituting variables, including functions and values, is computer science's generic atomic framework enforced by the object-oriented, capability-limited machine code of a Church-Turing Machine. The universal model of computation supports pure functional computations. It implements nature's clockworks privately for each instance of a species, a model without physical limitations devoid of undetected interference that scales equally for all, for eternity.

This formula privately matches form to function, and each living instance survives independently. Its decentralised model is essential for the colonization and civilisation of global cyberspace. The clockworks of the λ-calculus meta-machine frame the function abstractions as Application Threads in a secure DNA defined by a Namespace. The λ-calculus Variables, Functions, Abstractions, and Applications are bound on-demand, step-by-step, object-by-object, and frame-by-frame in a threaded, scientifically engineered, functional chain of atomic abstractions.

Capability Limited Addressing constrains all calculations, and the chained DNA strings of the namespace shape the end-to-end application as an evolvable yet stable interference-free instance of a defined species of dynamic software.

When infallibly automated, infinitely complicated DNA chains are woven functionally from the symbols on the virtual chalkboard of a namespace. The symbols link mathematics as scientific expressions and, using Capability Keys, the implementation can stretch safely and endlessly across universal cyberspace. At the same time, the mathematical chalkboard defines the directional DNA hierarchy for a private, individual computation controlled by a citizen of cyber society.

The Abacus and the Slide Rule implement the same core mathematical methods found in Babbage's Analytical Engine. These object-oriented machine functions are the core constituents of a Church-Turing Machine. When immutable Capability Keys as hardware assets guarantee Capability Limited Addressing to encapsulate every digital object, a fail-safe programmable machine of independent functional components is as dependable as Babbage's Analytical Engine's cogs, pumps, levers, and wheels. The hardware of Turing's α-machine acts as the survival machine to protect the software functions from corruption using hard digital shells to detect and prevent all sources of corruption from every attack vector in cyberspace. The result is Industrial Strength Computers, which apply the Babylonian Golden Rule to every instruction.

Function abstractions cover an infinite variety of algorithms used in computer science, but all software shares the same atomic form as defined by Turing's single-tape α-machine and the universal model of computation. Its atomic form materialises the infinite complexity of woven mathematics in science and nature. A clockwork λ-calculus meta-machine underpins Turing's α-machine controlling the technology of Capability Limited Addressing.

The result is profound. The laws of λ-calculus apply the technology of Capability Limited Addressing to the directional DNA hierarchy as a dynamic namespace security policy. The tangible Capability Keys direct this transparent security policy regulated by the citizens of cyberspace. While functions are public, data is private. The universal model of computation reaches into every corner of cyberspace. All abstractions are covered, securing the foundations to prevent Artificially Intelligent Software breakout, stop massive data breaches, the theft of blockchains and the corruption of 21st-century civilisation.

The abstractions are atomically equivalent, encapsulated as the symbols within the λ-calculus. When every frame of execution by an Abacus, in a Rose, or forming a Snowflake uses the universal model of computation, the form always matches the function. The universal model of computation pioneered by the Abacus is the natural model of stable evolution in a hostile environment. The Babylonian Golden Rules of the Abacus are safe for the future—the protected boundaries with dynamically guarded Ishtar Gates. Access-controlled gates are essential for the survival of digital software and a virtualised, hostile world that can thrive and prosper forever.

CHAPTER 6

Trusted Software

*'The world today is made, it is powered
by science; and for any man to abdicate
an interest in science is to walk with open
eyes towards slavery.'
J. Bronowski, from Science and Human
Values when at the Salk Institute for
Biological Studies, San Diego, CA, Feb
1964*

For decades, General-Purpose Computers prospered while ignoring the λ-calculus. Until now, computer scientists have ignored serious cybersecurity. It did not matter until democracy came under threat. However, the problems start with fragile, static machine code and unguarded memory sharing. Software cannot fix the hardware flaws of General-Purpose Computer Science.

Neither operating system nor patching works, and it matters. The Brave New World[68] of the information age cannot start with a flawed foundation. It will lead to digital dictatorship, as already experienced in Iran, Russia, China and North Korea. Babbage demonstrated

---

68   Brave New World (1932), best-known work of British writer Aldous Leonard Huxley, paints a grim picture of a scientifically organized utopia.

Industrial-Strength Computer Science at the height of the Industrial Revolution. Ada Lovelace went far further than her mentor. She used and recognized the power of symbolic functional programming to explain how to create music and art in addition to Babbage's tables of numbers. Only when the telecommunications industry demanded worldwide Industrial- Strength Computer Science the λ-calculus took its rightful place with object-oriented machine code.

The PP250 introduced this solution by focusing on trusted software, which requires detecting every software and hardware error and adding fail-safe recovery. Its approach copies Babbage's Analytical Engine, offering *Infallible Automation* to create flawless software through cybersecurity as Industrial Strength Computer Science.

Pioneered by Charles Babbage 200 years ago, Industrial Strength Computer Science builds on nature's universal model of computation using symbolic, functional variables. The power of the λ-calculus turns Alan Turing's simple binary computer into the digital gene of cyberspace. The atomic genes use digital guard rails, typed boundary shells, and the DNA of an application hierarchy as a species to keep dynamic computational threads on track.

Mechanical computers all strived for *Infallible Automation* while limited by that day's technology. It started with the Abacus, improved with the Slide Rule, and summited with Babbage's unfinished second engine.

The move to digital computers was different. The Church- Turing Thesis, circa 1936, introduced the λ-calculus and the Turing Machine. Still, the λ-calculus did not take off since cyberspace remained in the distant future, and programming was the immediate problem to solve. Then, on another track, high- density microelectronics became a significant distraction.

The nature of computer science changed from visible mechanics to invisible software and fantastic microelectronics, so bigger, faster, lighter, cheaper computer chips drove the mainframe software into the PC age.

Computer scientists ignored the λ-calculus, but the Church-Turing Thesis as the nature of computation remained and as Bronowski wrote, *'to abdicate an interest in science is to walk with open eyes towards slavery.'* He wrote these words after landing in Japan after WWII to witness the atomic destruction of life and cities.

Based on blind trust, who can doubt that General-Purpose Computer Science will lead to far worse results? Instead of impacting just two cities, the world will collapse when malware is faster than scientists. The unfairly privileged General-Purpose Computers is a corrupt cyber dictator. These machines will lead civilised society to branded digital dictatorship, corruption on a global scale, and digital slavery. Securing cyberspace is vital for survival as a civilised, productive, democratic nation. In an unending future, purging the operating system's hardware privileges to eliminate malware is essential for the benefit of all.

The virtualization of society demands equality, freedom, and justice for all. It is paramount. It is impossible for General-Purpose Computers—the most severe shortcoming is the growth of global monopolies as centralized digital dictators. Displacing democracy with centralized dictatorship is retrogressive.

Global monopolies are building self-serving digital dictatorships. A global market space that farms governments and citizens like flocks of sheep using malware as their sheepdog. They may respond to national regulations, but suppliers dictate the rules.

The problems began with a lack of competition ever since the birth of the Intel 8080. Since then, innovations like PP250 have stalled, and backward compatibility is the 'name of the game.' Without competition, corrupt dictatorships and enemy states will destroy what remains of civilised democracy.

The end game of cyberspace must preserve, protect, sustain, and defend freedom, equality, justice, and a government of the people, by and for the people. Cyberspace must protect the cornerstones of civilisation and prevent centralized dictators.

Babbage's hallmark of Infallible Automation is fundamental for trusted software and fail-safe computer science. Industrial Strength Computer Science that applies the λ-calculus is the only safe solution for the future of democracy. Both sides of the Church-Turing Thesis are needed to scientifically level the digital playing field and balance the needs of citizens and businesses through trusted software and fail-safe hardware. Trusted software protects society from every human mistake in building and extending the most complicated of all machines.

The Church-Turing Machine reverses the unfair centralized power of suppliers, corrupt governments, criminals, enemies, and self-serving monopolies. Instead, cyber power in capability keys always belongs to individuals. Once cyber security and confidentiality are in the hands of the citizens, democracy results, and dictators are avoided.

Infallible automation scientifically levels the playing field of cyberspace as a trusted platform for safe and secure use by any citizen as a cyber democracy. Using a λ-calculus meta-machine with capability-limited addressing deconstructs the monolithic centralization of General-Purpose Computer Science, distributing powers symbolically to individual citizens with private needs or an approved or elected responsibility.

It was in 1965 when Jack Dennis and Earl van Horn published the seminal paper on Capability Limited Addressing *'Programming Semantics for Multiprogrammed Computations.'* Bob Fabry explored the idea in his doctoral dissertation on the (Chicago) *Magic Number Machine*[69]. Maurice Wilkes was enchanted; he took the idea to the Plessey team as the foundation technology for telecommunications, and the Industrial Strength Computer called PP250 resulted. Soon after, he started the CAP at Cambridge University as a software research project. However, only the PP250 removed all centralized threats and superuser privileges.

---

69   R. S. Fabry. List-Structured Addressing. Ph.D. thesis, University of Chicago, March 1971

By 1972, the PP250 was in service and became well-known as the archetype of Capability-Based Computers. In 1976, Peter Denning asked me to write a book on PP250 for his Operating Systems series. However, while PP250 was a successful prototype for *Infallible Automation* and fault-tolerant software, it rejected the accepted theory on operating systems. Operating systems and monolithic virtual machines have no place in Industrial Strength Computer Science. Digital convergence was unrecognised, and the whole story was too early to tell.

TRUSTED SOFTWARE

The pre-electronic age was outdated by digital convergence. Blind trust in software and shared memory with unfairly privileged administrators that includes a centralised operating system and unelected, networked superusers misfires in the global electronic village of a digitally converged, electronic age world. The mantra for *Infallible Automation* in the electronic age must be democratic. *'Humans cannot be trusted, and neither can their Software.'* The solution demands distributed ownership of functional objects secured by digital boundaries to guarantee the functions of the software according to a need to know with the least authority.

The only way to avoid the pitfalls of General-Purpose Computer Science is to scale, naturally, from small atomic functions to the endless, expanding universe of cyberspace—the universal model of computation and the Church-Turing Thesis team with the λ-calculus to solve all computer problems. A Church-Turing Machine precisely addresses the required science for a future safe cyberspace.

Trusted software depends on more than protected functional boundaries. The structure of the function matters as much. While digital boundaries are fixed, the structure must be program-controlled to match the ever-changing framework and forms of the related functions. Capability Limited Addressing began as a hardware technology to limit software to match the function. Any connection with the λ-calculus was informal.

The Church-Turing Thesis formalises the relationship as a λ-calculus meta-machine. The λ-calculus meta-machine is a Church-

Machine. Like the Turing Machine, it is program- controlled, but instead of imperative Turing Commands, functional Church Instruction navigate the directional DNA hierarchy of an application. Encapsulating Turing's single tape α- machine in a λ-calculus meta-machine applies Alonzo Church's mathematical formula to constrain the unruly Turing Commands and liberate static binding with the powers of the λ-calculus. Embedding the λ-calculus into a Church-Turing Machine not only levels the playing field between software and hardware, but the dynamic powers of the λ-calculus and functional programming are available in machine code to reach worldwide.

When reduced by the laws of the λ-calculus, mathematical symbols are, by nature, modular. They obey the universal model of computation as object-oriented machine code, which can now extend and communicate reliably and globally. A Capability- Based meta-machine is the same clockwork framework started by the Abacus that guarantees the boundaries and structures relationships as the universal model of computation, applying the laws of the λ-calculus continuously and seamlessly as in nature.

The symbols in mathematics are trusted atomic components in a calculation; as capability-limited digital assemblies, they form a robust, fully functional, dependable namespace expressed on a chalkboard at school or university. The symbols are implemented and engineered to meet a calibrated MTBF for a qualified level of trusted performance. The symbols exist as a software-defined machine's hardened, measured components. The object- oriented machine code is encapsulated and structured by the immutable Capability Keys grouped in C-Lists as strings of mathematical DNA. Function abstractions connect and communicate universally, point to point, through communications servants either synchronously or asynchronously. The unquestioningly trusted monolithic software and downloads from uncertain third parties using unfair hardware privileges evaporate, replaced by hardened mathematics with a calibrated, published MTBF that can keep us safe as we colonise cyberspace. Calibrated crash testing modular function abstractions will innovate the software marketplace as crash testing, and airbags did for the car industry.

Reconsider the Abacus. It is a mathematical machine. Just two programmed function abstractions exist stored in the human mind. In a Church-Turing Machine, the wooden framework, replaced by the directional DNA hierarchy of Capability Keys, provides atomic security that can't be bypassed. When a Church Instruction unlocks a Capability Key, the object access rights are bound to a capability context register. Each abstraction is sheltered as a private computation frame, defined by a set of private variables implemented as immutable capability tokens.

No centralised operating system or unnatural privileges complicate the dynamically framed and privately protected atomic calculations. Any functional need provided this way, as first proved by the arrival of the Slide Rule and the Difference Engine, uses the λ-calculus, Capability Limited Addressing guarantees. The modular function abstractions become hardened functions implemented as object-oriented machine code. A Church-Turing Machine implements the enduring results from the same expressions learned at school.

In a network, this architecture can spawn remote threads that work in parallel without any loss and with the potential for significant gains. The telecommunications application posed the perfect example for secure and safe parallel network distribution, functions unavailable with monolithic software and statically bound procedures. Industrial Strength Computer Science is nuclear, emphasising qualified software as reliably calibrated components. A Church-Turing Machine addresses hardware and software in a unified way to engineer trusted standards of functional reliability rigorously.

The monolithic compilations in General-Purpose Computers are stupidly unchecked as the rubber hits the road on the wild frontier of computer science. Containing this explosive force of nature is like engineering any powerful machine. With every cycle of the pistons, any explosive power must be contained and channelled to maximise performance, minimise loss, and prevent a human catastrophe. Likewise, the λ-calculus meta-machine encapsulates the machine cycles of explosive instructions performed by Turing's α-machine. The dangerous results of each cycle are regulated and contained by a Church-Turing Machine at the trigger point of ignition.

While mathematics scale scientifically, it was inevitable that branded centralised monolithic software and General-Purpose Computer Science could never grow as an engineered fail-safe global network. Malware and hacking prove this point, and adding fragmented linear memory increases the networking threats. The imprudent first mistake was unquestioningly sharing all access rights within a monolithic virtual machine. It worked while mainframes ran in sealed rooms for eight-hour shifts under devoted, expert supervision. It cannot work for a continuously dependent society of busy civilians engaged in their problems or, most significantly, protect the cornerstones of civilian democracy.

Capability Limited Addressing and object-oriented machine code encapsulate programs on the surface of computer science without all the highly technical baggage that places civilians at risk. More than this, Capability Keys safely connect the surface activity of computer science to the hands of the citizens of cyber society and the centralised dictatorial threat to democracy is resolved. Confidential data remains private and detailed in- depth. The universal model of computation regulates healthy rules of delegated ownership and encapsulated executions. Power maximised by the laws of the λ-calculus splits confidential data from public functions. The independently sourced information objects are separated as protected symbols with different security needs and separately managed security policies.

A hybrid choice for Capability Limited Addressing remains fatally flawed by the shared assumption of von Neuman after WWII and Butler Lampson during the Cold War. The pre- electronic age still infects CHERI, the latest capability-based computer from Cambridge University. These backward- compatible machines cannot support object-oriented machine code. They lack any Church Instructions to bind data to mathematical symbols instead of shared pages of linear memory. Thus, these computers remain tarnished by the baggage of General-Purpose Computer Science.

Industrial Strength Computer Science maximises results at the point and time when software touches hardware. Nothing less serves the everlasting future of sleeping with the dark side of humanity corrupting cyberspace.

General-purpose computers are compromised and interfere with the progress of digital civilisation. Already, the loss of trust in the national government has increased. The impact on society from digital fraud, crafted forgery, fake news, deep fakes, election interference, industrial espionage, sabotage, and corruption frightens innocent civil society. As a WMD[70], General- Purpose Computer Science has a price to pay. The real story driving Industrial Strength Computer Science is saving democracy from dictatorship, decay, and downfall.

The importance of this story repeats America's fight for independence and the greater good of freedom, equality, and justice. Mixing AI malware with virtualised society in General-Purpose Computer Science only leads to the dark side of LIFE 3.0, foretold by Max Tegmark. Trusted software and fail-safe hardware are essential to the nation's progress and prevent the dark side from taking control. Trusted software must become the general case for conflicted cyber-nations and rogue states to co- exist in the digitally converged electronic village of the 21st century and beyond.

Circa 1972, when PP250 launched Industrial Strength Computer Science for global telecommunications, the primary threat was undetected, self-generated errors. Today, the risk is Artificially Intelligent Malware used by criminals and enemies. The danger of malware and privileged hacking is already formidable and poorly understood when fixed by an urgent patched upgrade. Regression testing is expensive and incomplete. Increasingly, as the virtualization of society grows, monolithic software becomes too hard to evaluate. The difficulty with relaunching the 737 MAX demonstrates this point. The need for direct pilot control over the software that virtualized the plane's centre of gravity accentuated the flaw. The monolithic General-Purpose Computer software in a life-supporting application is too complex to evaluate and too centralized as a dictator.

The move to software dictatorship alone justifies the move to the Church-Turing Machine as the essential next step in the progress of computer science. The first papers published on the fault-tolerance of PP250 showed that 80% of software errors were detected within

---

70   Weapon of Mass Destruction discussed in Book 2 The Fate of AI Society

20% of a code block execution, and 100% of errors were detected once all paths in the block were evaluated[71]. These results allowed PP250 to eliminate dormant bug problems, malevolent software, and the superuser administrator.

A.I. software is unstoppable in specific situations, and patched upgrades are insufficient if super-user privileges are hacked remotely. Worse still, subcontracting the almighty power of administrators to some of the most corrupt, criminal, dangerous, low-cost places on earth gives direct access to critical industrial services. Adding A.I. to the mix will increase the destructive power of state enemies in unexpected ways.

The efforts to add cybersecurity on top of flawed computers, as reviewed at the end of this Chapter[72], but in every case except a Church-Turing Machine software solution starting with Identity Based Access Control, still leave all the unavoidable privileges and unacceptable threats of a remote hacked superuser account.

With the lack of modular error detection, these unfair privileges are holding back progress and, unless removed, will one day destroy democracy. The mainstream takes no notice. The vested interests of monopolies enslave decision-makers, and over- involved spy agencies for homeland security may never be convinced to give up their unconstitutional advantages. It requires a national effort like the Manhattan Project to win another, but this time unseen world war. The digital dictators only offer patched upgrades and fragmented workarounds that do not solve the underlying problems.

---

71   R. J. Leaman and M. H. Lloyd and C. S. Repton. The development and testing of a processor self-test program pp 308—314, The Computer Journal Volume 16, Number 4, November 1973, reproduced at sipantic.com blog August 2019.

72   Object-Capability software exists in a wide range from ACLs as columns reference or table rows, capabilities as keys or as objects in a variety of packages including Unix fs, setfacl(), NT acls, POSIX, SPKI, Unix file descriptors, Hydra, keykos, W7, EROS, and E language, but they all share a General- Purpose Computer with unavoidable, undetected errors and unfair privileges.

FAITHFULL SOFTWARE

What is not yet recognised but is far more shocking is the impending destruction of any virtualised democracy. Industrial Strength Computer Science is more than a search for a scientific computer. It is a fight for the future of democracy. A global cyberwar is underway between good and evil, and the survival of America's government of the people, by the people, and for the people is at stake. More than a search for cybersecurity, it is the search for a future-safe digital platform to serve as a secure virtualised citizen's government in a digitally converged, virtualised world.

The endless future of society will unavoidably be digitally converged. In this new world, laws change, opinions conflict, the news is made up, commercial incentives are transnational, criminal gangs are global, international enemies can sabotage, fraud and forgery are digitised, malware and hackers hide, and dictatorships thrive. An already fragile world run as an unregulated global village founded on corrupt digitally converged electronics. Industrial Strength Computer Science and trusted software are the only way to civilise the digital paradigm that now runs society.

Rigid, functional boundaries that isolate endlessly complicated functions on private virtual chalkboards are the high ground if cyberspace is to meet the needs of a democratic cyber society in this conflicted, digitally converged world. The Church- Turing Machine is the only computer to meet this need. Industrial Strength Computer

Science must be recognised as a national emergency long before AI enslaves us all. Removing blind trust and default privileges is the first step—the replacement must be the clockwork cybersecurity of a λ-calculus meta-machine.

The minimal transistor count for a λ-calculus meta-machine can be qualified and proven once to mass-produce trusted, modular software that needs no privileged baggage. The Church- Turing Machine uses object-oriented machine code and is a scientific architecture, guaranteed both as fail-safe and future- safe. The qualified reliability characteristics can be published just as fuel consumption, crash testing and energy efficiency are documented and published for other industries.

However, the resistance to change is strong. SOSP'77[73] reviewed Capability Limited Computers by a panel of experts, including the author, led by Bob Fabry, who was chairman. It was one-sided and remained that way. The computer experts promptly dismissed PP250 as an abnormal case. The experts sold the day's hype that secure software kernels would work. The fairy tale crashed in the ashes of Microsoft Vista while Steve Jobs used Objective-C and the iPhone, making Butler's *'abnormal case'* the general case for digital convergence.

This reversal of fortunes in the status of General-Purpose Computer Science makes Capability Limited Addressing and the Church-Turing Machine a vital necessity. Trusted software is mass-produced dynamically by the λ-calculus in a Church-Turing machine, as this architecture has already proven. Indeed, the Church-Turing Machine is essential for a virtualised society's future and survival. Only a Church-Turing Machine can mass produce infallibly automated software to the calibrated grade of service acceptable for national

---

73   The Symposium on Operating Systems Principles (SOSP), organized by the Association for Computing Machinery (ACM), is the most prestigious single-track academic conferences on operating systems, Symposium on Operating Systems Principles - Wikipedia

security and democratic society. Life-supporting services run by *'Infallible Automation'* are essential for a digitally converged world. Cybercrime is the fastest-growing industry in the world. The opportunity for cybercrime expands dramatically as the Internet of Things grows. Using an out-of-date computer architecture from WWII, the rush forward becomes suicidal.

TOO COMPLEX TO TEST

This dangerous situation demands more than von Neumann's WWII architecture and Identity-Limited Access Control policy can offer. The process of digital convergence is not just a shared global platform; the software is life-supporting, virtualizing every aspect of human activity, from birth to death, in industrial society. The survival of the nation depends on trusted software, mass-produced on demand. Centralized software security is a failure. It gets far too complex to test,

Instead, it must be incremental and distributed, held by the individual users of cyberspace who guide the fail-safe services that stretch across the globe. This faithful machine is a trusted utility, profoundly vital to a free society, democratic nations, and the progress of civilisation.

Misled by the pre-electronic age dream of a foolproof Kernel and Cold War centralization, General-Purpose Computer Science catastrophically fails to serve either mathematical science or democratic society. General-Purpose Computer Science beguiles the public but does not secure the future because the software is unfaithful. As measured by upgrades, the failure rate is months instead of decades.

Life-supporting services like the 737 MAX demand years. The complexity of monolithic General-Purpose Computer Science has grown too complex and centralized to evaluate. Software services, when fielded with a flaw or targeted and ruthlessly attacked, inevitably crash.

As an industry, computer science continues to follow a dangerous road downhill. The A.I. researcher Pedro Domingos notes in his book *The Master Algorithm*, *'People worry that computers will get too smart and take over the world, but the real problem is that they're too stupid, and they've already taken over the world.'*

Mainstream resistance was the problem at SOSP'77 when batch processing and timesharing kernels took over the computer industry. That choice created problems for every line of software that exists today. The challenges include A.I. breakout, blockchain theft, Facebook, Google, and every software application hoarding confidential data that rightly belongs to the citizens.

Serious problems exist all created by the General-Purpose Computer:

- On the spot, error detection is absent.

- Inadequate Identity-Based Access Control

- Backdoor attacks.

- Exhaustive testing in field conditions is impossible.

- Monolithic compilations fail in untested ways.

- Skilled staff shortages exist.

- Patched upgrades are too little and too late.

- Civilians cannot cope with designs made by experts.

Increasing complexity makes life in the 21st century impossible. General-purpose computers are like the piles of dysfunctional stones rejected by the Babylonian marketplace. No matter how many patched upgrades are released, it will not get better. The problems are systemic; they feed off one another.

CRIME PAYS, DICTATORS RULE

The inevitable consequence of digital convergence is already advanced, and the end games between two polar alternatives are in view. Plessey Telecommunication faced this choice with the first attempts to computerise the phone network. The unavoidable impact on society started long before Capability Limited Addressing and object-oriented machine code ruggedised the PP250. The easy commercial alternative adopted in WWII led downhill to endless individual, social, national and international catastrophes.

Global cyberspace is not the same problem as isolated batch processing, just as a Boeing 737 MAX differs from a WWII Spitfire. Technology and requirements turned life upside down. The Spitfire won the Battle of Britain, and the General-Purpose Computer began the information age, but both are outdated. *Infallible Automation* under client supervision is required in cyberspace. Times change, and best practices and patched upgrades fail with life-supporting applications in a digitally converged society. Even small failures mean people will die. National survival cannot depend on the best practices of unregulated General-Purpose Computer Science. Applications must be qualified to deal with worst-case, real-time conditions; a batch processor in a locked room full of experts is trivial compared to non-stop, life-supporting software in the globally converged digital village.

The dictatorial nature of the General-Purpose Computer is fighting with society at the worst moments. Consider the pilot and copilot fighting for control in the cockpit of the Boeing 737 MAX aircraft. The software won, and two planes crashed vertically into the earth, killing over three hundred innocent civilians and grounding the aircraft indefinitely. Society cannot trust Computers run by centralized operating systems. Their intrusions into democratic, industrial society are fatal. Software errors must be detected and resolved on the spot, on the wild digital frontier where software touches hardware, expending energy.

The lack of industrial strength, digital power's centralisation, and compiled software's short half-life are debilitating—the dichotomy amplified by human incompetence, criminal gangs, and global disparities only leads to conflict. While citizens deserve privacy, monopolies hoard their data for sale.

The conflicted world is at war in unguarded cyberspace. The fight starts on the digital frontier, at the atomic level, then festers and grows to an unpleasant, unavoidable end. Cybersecurity can only be detected and resolved atomically. It is not a choice. It is essential for a level playing field where law and order frame trusted software throughout cyberspace.

On paper, American democracy and criminal dictatorships are polarised forms of Government, but in cyberspace, they both spy to sustain their domination. Citizens' rights in cyberspace will end if virtualization proceeds this way. The dangers in General- Purpose Computer Science override the institutions of good government. Best practices and branded dictatorial platforms are incompatible with constitutional democracy, transparent law and order, and the traditions of precedence.

Nevertheless, virtualization is rapidly proceeding but will not sustain a government of the people, by the people, and for the people. General-purpose computer science is unacceptable for public use. The playing field is tilted to favour experts, suppliers, criminals, and dictators. Law, order, and justice fail because unfair privileges overwhelm the written constitution. Unelected dictators build transnational digital kingdoms like TikTok, Facebook, Google, Microsoft, and others that place global domination above constitutional democracy. The message that drives cyberspace is *crime pays, and dictators rule.* It is antithetical to civilised society.

THE ROAD TO CYBER-SECURITY

Industries already spend a fortune on cybersecurity using technology that cannot guarantee results. Cybersecurity is considered the biggest threat to prosperity over the next decade. Three million unfilled cybersecurity positions exist worldwide, and the tenure for a chief information security officer (CISO) has fallen to less than two years[74]. The constant 7x24 hours of stress, the urgent panic of the job, the unsurmountable challenges and the shortage of qualified staff are the reasons. A report by the Ponemon Institute found that it takes a year to discover a severe data breach, and 65% of IT and security professionals quit because of burnout[75].

Whatever is tried or spent, the problems remain. Crime pays because General-Purpose Computers ignore fundamental error detection, and things will only get worse. The Future of Life Institute warns that cyber civilisation is self-destructive, and ever-smarter software will break out and take control. It makes the scientific solution to cybersecurity a national priority.

Circa 1936, the hardware and the software discoveries of Alan Turing and Alonzo Church framed the Church-Turing Thesis, which

---

74   CNBC News https://www.cnbc.com/2019/10/11/65percent-of-stressed-out-cybersecurity-it-workers-think-about-quitting.html

75   IBM Data Breach Study: Impact of Business Continuity Management Benchmark research sponsored by IBM Independently conducted by Ponemon Institute LLC October 2018

explains both sides of computer science. The λ-calculus is the scientific glue that binds the logic to the physics. A General- Purpose Computer only supports the physical hardware; von Neumann ignored the software requirements that depend on the laws of the λ-calculus. Consequently, the foundation of mathematical software and the universal model of computation defined by nature and history are missing.

Industrial Strength Computer Science requires a λ-calculus meta-machine to frame the basement technology of digital society. The λ-calculus as a Church-Turing Machine harnesses binary data as powerful digital objects. The check and balance create the universal model of computation as a flawless and endless network of infallibly automated mathematical, functional science.

A λ-calculus meta-machine levels the playing field of computer science. These rules are universal and easy to use. Universally fair mathematics from the universal model of computation replicates the petals of a rose, the hexagonal honeycomb bees used to survive and the decimal rails of the abacus. The strings of DNA stabilise evolving functions as different species of life. In this world, the universal model of computation smooths the operation and evolution of software survival in hostile cyberspace.

The contested results of the 2016 Presidential elections prove the danger of corruption through a clickjacking attack, and the trouble is only beginning. To prevent attacks from succeeding, replacing General-Purpose Computers is vital. The General- Purpose Computer cannot protect themselves, so they certainly cannot protect cyber-dependent societies of the future.

Security for the Internet of Things is considered a nightmare, so starting here makes the conversion process meaningful and authentic. It is a fertile, greenfield opportunity to engage the future-safe, cyber-secure alternative of Church-Turing Machines.

Any government regulation, such as the breakup of software monopolies, will fail unless the dangerous hardware from computer suppliers ends. The lack of hardware architectural competition is at fault.

Given competition, the Church-Turing Thesis leads to trusted commercial computers and an Internet upgrade with digital security. Much has already been achieved by software, as shown in Figure 8. Given the availability of microprocessors that implement the Church-Turing Machine discussed in Figure 7.

A CHURCH-TURING MACHINE

ook at the difference between the technology stacks of a General-Purpose Computer (on the left) and a Church-Turing Machine (on the right) in Figure 7. The shaded area shows the hardware part. The right side is more straightforward and has all the features of the Church-Turing Thesis that reduce the complexity and overhead of General-Purpose Computers. A better form of computer science begins with the λ-calculus, supported by a Church-Turing machine to wrap the binary computer, using Capability Limited Addressing. The wrapped object-oriented machine code safeguards the mathematics of Turing's single tape α-machine as a dynamically constrained binary computer.

Untrusted Software	Trusted Computer Science
Isolated Virtual Machines	Encapsulated Chalkboards
Programming Languages	λ-Mathematics (Namespace)
Image Compilers	λ-Applications (Threads)
Operating Systems	λ-Function Abstractions
Timeshare Virtual Memory	Object-Oriented Machine Code
Shared Memory	Turing's α-machine
Imperative Commands	Functional Church Instructions
Over-Stretched Turing Machine	Clockwork λ-calculus Meta-Machine

FIGURE 7: TECHNOLOGY STACKS OF GENERAL-PURPOSE COMPUTERS VS CHURCH- TURING MACHINES

Monolithic virtual machines become individual functional objects. The change enables discrete units of computation accessed by Capability Keys using the universal model of computation. Each unit is one object-oriented function that runs in a Thread. Capability-limited addressing with the universal model of computation protects private data and controls the functions of computer science as linked DNA strings and secure DNA chains in an application Namespace.

Like tangible physical assets, Capability Keys are privately created and returned accordingly. It civilises the deconstruction of monolithic, dictatorial powers in cyberspace. Capability Keys are immutable digital tokens that encapsulate and hide each digital object's binary content from interference and misuse— the encapsulated objects represent one symbol in a scientific expression from a virtual chalkboard. The Namespace hierarchy defines the DNA hierarchy of a software application as a fail-safe machine, all privately controlled by the citizens in a cyber society.

As digital assets, Capability Keys are private until shared. The capability-enforced atomic modularity dismantles a General-purpose computer's centralized and autocratic forces.

For example, when making a digital copy of a human identity with facial recognition, the digital reference needs a 'stamp' to maintain the correctness of digital information and quickly detect a deep fake. Capability keys act as a stamp to track back to the trustworthy source through the directional DNA hierarchy. In the simplest form, a Capability Key (in the DNA) has a true or false indicator that shows if the digital data has changed. The stamp alerts the user to an altered object, where an object can be a complex abstraction, giving citizens control of private data and safeguarding their identity from deep fakes.

The centralised General-Purpose Computer passes data to the wrong hands, and the tools for faithful watermarks are missing. These tools must operate on the surface of computer science where and when physical change occurs. In a Church-Turing Machine, watermarks trace ownership through the private DNA chains of each Namespace by preserving identity and digital ownership of data on the digital frontier where all changes begin. Capability Keys correctly separate and distribute power between services and data, limiting both data and services for use by approved agents. In the digital world of the 21st century, individuals and society must remain in control of their intellectual property. It is critical to the correct functioning and the survival of a citizen's cyber-democracy.

At the same time, compiled images and blockchains cannot alone guarantee results. Without capabilities, the hands of enemies and conflicted nations will use Artificially Intelligent Malware to break out unlimited attacks. Ransomware attacks prove Identity-Based Access Control is inadequate. The theft or encryption of a blockchain using ransomware is real. The harm caused by such formidable attacks is unpredictable, and the increasing degree of catastrophe is unavoidable. If Artificially Intelligent Malware rules cyberspace, the future of Life 3.0 is grim.

Throughout civilised history, democracy emerged to counter human weakness, centralised power, and the default privileges of dictatorship. Democracy term limits delegated power as decided by the people through elections. Titles and insignia define power in a society, from a meeting Chairperson to a nation's President.

Capability keys must exist for cyberspace to be democratic instead of dictatorial, and cyberspace cannot be oppressive. Tokens began as chains of office, the symbol of power. In a democracy, the delegation of power is approved by citizens. These mechanisms of democracy can exist as a protected digital abstraction, guaranteed mathematically in a Church-Turing Machine. These tokens of power are the immutable Capability Keys to functional abstractions representing the institutions of good government, but only in a Church-Turing Machine can this be guaranteed. Capability Keys approve power in all forms, which can also be withdrawn. The Capability Keys of a Church-Turing Machine provide the means to democratise cyber society.

OBJECT-CAPABILITY SYSTEMS

In mathematical cyberspace, symbols like Capability Keys define power and authority, but only a Church-Turing Machine connects this power from the digital frontier to the hands of the citizen. It is the only form of cyber power that can be withdrawn. The only form of power that can be delegated to another for a term limit, even removed when approved. Digital tokens are the gold of cyberspace. It is the essential substance to civilize cyberspace.

Property	Software Systems on General-Purpose Computer				Hardware Systems	
	Model 1	Model 2	Model 3	Model 4	Model 5	Model 6
	ACLs as columns	Capabilities as rows	Capabilities as keys	Object capabilities	Hybrid hardware	Meta-machine
Example Software	Unix fs, setfacl(), NTACLs	POSIX	SPKI Unix file descriptors	Hydra, keykos, EROS,E	CAP, CHERI	PP250 Church-Turing Machines
Hardware	General-purpose computer				+ extension	
A. No Designation Without Authority		Unspecified (possible)	No			
B. Dynamic Subject Creation		Models 2, 3 & 4				
C. Subject-Aggregated Authority	No		Model 3 & 4	Model 4	Yes	
D. No Ambient Authority		No				Yes
E. Authority Composability	Unspecified		No		Unspecified	
F. Access-Controlled Delegation					Yes	
G. Dynamic Resource Creation	All Models					

H. No Shared Memory		
I. No Superuser or Default rings of Privileges	False	True
J. λ- Meta-machine, confidential data, and the universal model of computation	True	

FIGURE 8: COMPARISON OF CYBER SECURITY MODELS

Artificially Intelligent software must be a servant, not a master—however, a servant of the people, not a dictator. Life in the 21st century will not deteriorate if Industrial Strength Computer Science dominates because democracy will flourish.

The advantages have increasingly been understood and appreciated since the PP250 and CAP machines of the 1970s. Computer security interests have improved from the Cold War Identity Based Access Control standard, as shown in Figure 8: Comparison of Cyber Security Models. From left to right, shows improving alternatives. The first four are software-only solutions using general-purpose computers, while Models 5 and 6 use capability hardware, first in a hybrid solution and ultimately, like PP250, without any centralized operating system.

- Model 5 is a hybrid solution that adds segmented memory protection. It obscures the λ-calculus and blocks networked capability mechanisms by reusing the centralized operating system of a General- Purpose Computer.

- Model 6 is a Church-Turing Machine that uses transparent Church Instructions to unify the mathematics of Capability Keys to reach across cyberspace without interference.

The steady acceptance of object-oriented programming and Capability Limited Addressing progress towards a Church-Turing Machine Architecture, shown on the extreme right as Model 6 in Figure 8. The change reduces the operating system to a handful of Church Instructions. The solution sidesteps the unsolvable von Neumann assumptions, removes the centralised baggage and reduces total software size and cost while increasing quality, functionality, and performance.

Instead of overstretching the Turing machine, the scientific foundation of computer science is established as a clockwork λ-calculus meta-machine.

Software solutions use digital capability keys for all the right reasons, but they still depend on the default superuser privileges of General-Purpose computer science and identity-based access control. The extreme Object-Capability software systems, shown on the right as Model 6 in Figure 8, represent a Church-Turing Machine.

The chart is revised from the work of Miller et al. in 'Capability Myths Demolished' to correct and clarify CAP, CHERI and add PP250 as a Church-Turing Machine. Miller et al. explain why object capabilities matter and why Access Control Lists (ACL) fail. Still, this paper does not discuss the underlying problems of General-Purpose Computers or include the ultimate endgame of a Church-Turing Machine added as Model 6.

Capability Myths Demolished only compares software security models[76] starting on the left with Butler Lampson's original Identity Based Access Control, ACL model from the Cold War era. Moving to the right are improved but seldom used software-capability-based operating system solutions. These intermediate steps signal the progressive acceptance of object- oriented software and capability-based security over the prior alternatives.

On the right is the extreme Church-Turing Machine represented by PP250. PP250 is the only computer that uses object-oriented machine code and walks away from all the dangerous flaws of a centralized operating system. Constantly ignored by the mainstream, only PP250 can be called a Church- Turing Machine. CAP and CHERI have no meta-machine or functional Church Instructions. They still use opaque controls and centralised, privileged software on

---

76   Capability Myths Demolished Miller et al, see http://srl.cs.jhu.edu/pubs/SRL2003-02.pdf

traditional General-Purpose Computers. All the attendant baggage and flaws infect these alternatives. The hackable superuser account remains. No matter how perfect the software is, the programs remain vulnerable to undetected attacks like Ransomware. They stay in place to attack the future.

The Church-Turing Machine is the exception, but the lack of competition in solving cybersecurity leaves PP250 as the only commercial example. Model 6 is the only approach where citizen users own their data in a private λ-calculus namespace. Model 6 is the only column supporting every Miller requirement as an expanded matrix. Model 6 is the only solution that replaces shared memory and default privileges with a λ-calculus meta- machine to seal the data leaks in the basement of computer science.

Those interested in more technical details are encouraged to read *'Capability Myths Demolished'* referenced above. This short paper covers the arguments that first led to the choice of extreme object-oriented machine code for PP250 using Capability Limited Addressing. The impressive results of the PP250 prove the well-established arguments taken for granted in this book. This solution drops all the baggage and operates directly on the digital frontier of computer science. It is impossible to bypass and engineered to guarantee Babbage's flawless, *Infallible Automation*.

A Church-Turing Machine is the only solution acceptable for a civilised democratic society. All the other models depend on von Neumann's flawed assumptions. General-purpose computers will always fail the tests of time. The assumptions from the pre-electronic age and the Cold War defined the almighty superuser and the centralised operating system. Still, they only increased the threat of von Neumann's shared memory architecture, sustained by commercial monopoly interests that prevent open competition through backward compatibility.

Using object-oriented machine code hidden by functional Church Instructions empowers the seamless, universal model of computation with transparent, program-controlled security policies that support a people's democracy in cyberspace: Capability Keys and the laws of λ-calculus place policy control in the hands of the users. Now, individuals control their confidential data security. Transferring ownership of confidential data to the people is the only way to guarantee a computer-dependent cyber democracy. General-purpose computer science infects life- supporting software, harming the nation in the digitally converged 21st century. Enemies share the same digital space in the same shared electronic village.

Digitally converged cyberspace cannot survive, driven by pre-electronic age standards of centralised control and enslaved citizens. The nation operating on a lawless digital frontier has surrendered law, order, and justice. A citizen's government of the people does not exist because the level playing field is missing.

Spies, monopolies, criminals, and enemies thrive, and the future tilts against law and order in civilized society. The Church- Turing Thesis holds the answer. While General-Purpose Computer Science slopes downhill into cybercrime, digital corruption, and dictatorships, the Church-Turing Machine defines the demanding grade of service for eternally trusted software needed for an endless life in Civilised Cyberspace.

CHAPTER 7

Trusted Computers

*If debugging is the process of removing
software bugs, then programming must
be the process of putting them in.
Edsger Wybe Dijkstra, 1930 – 2002, was
a Dutch scientist, programmer, software
engineer, essayist, and pioneer in
computing science.*

As the archetype of Industrial Strength Computer Science, the Capability-Based PP250 pioneered a rigorous form of the Church- Turing Thesis using the universal model of computation, the λ- calculus, and Capability Limited Addressing. A λ-calculus meta- machine adds a handful of Church Instructions designed for functional modularity and cybersecurity, using Capability Keys to access a handful of well-protected λ-calculus machine types defined in capability context registers. The fail-safe design is resilient to malware and hackers. Capability Limited Addressing detects and isolates program bugs, instruction by instruction, in a machine cycle that removes the superuser and all the default privileges of monolithic operating systems, leaving behind only pure functions like mathematics and logic.

Babbage's standard of *Infallible Automation* emerges from the universal model of computation that reverses von Neumann's assumption of blind trust in software quality. The mantra for Industrial Strength Computer Science is *'software- cannot-be-trusted.'* A Church-Turing Machine is designed to detect software or hardware failures using the orthogonal check and balance of Capability Limited Addressing applied mathematically, logically, and continuously on the spot. This spot is in the cockpit of a digital computer as individual machine instructions fire. The cockpit is the spot when errors can be found most easily, as software and hardware touch.

For example, the application team approves the Namespace assembly in advance to avoid malware. Like the airline industry after 9/11, terrorists are locked out of the PP250 cockpit.

When the computer revolution began, hardware remained bulky and costly. The PP250 computer used a 25-bit word length and occupied eight cubic feet. About 1000 Texas Instruments 7400-Series TTL components housed the machine registers and microcode of the Capability Limited Computer.

Half the hardware was the Church-Machine, including the capability context registers and the handful of Church Instruction. About fifty out of 100 micro-coded steps were for the λ-calculus meta-machine, and the remaining half of the PP250 was a context-limited, unstretched binary computer for the RISC instructions of Turings's α-machine. The micro-steps of the object-oriented machine code are summarized in Figure 9.

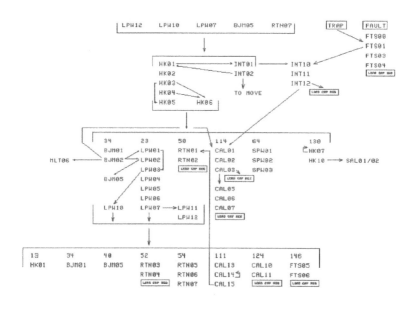

FIGURE 9: PP250 MICRO CODE LINKAGE STEPS

An extreme Church-Turing Machine removes all von Neumann's assumptions, artefacts, and side effects. The shared virtual memory, the central operating system, the privileged superuser, and the time-shared virtual machines evaporate, the proprietary baggage of a bygone age. Instead, the λ-calculus in the Church-Turing Thesis navigates the metadata held by the Namespace from the object-oriented assembly process across a level computational playing field.

In addition, the PP250 was a symmetrical multiprocessor—a reliable group of λ-calculus meta-machines operating together, accessing the same networked memory space. They could run for decades without interruption, continuous service without any downside like the curse of hidden malware, a data breach, software breakout, disastrous event, unnoticed bug, or a loss or compromise of confidential data.

The extended life span of software dramatically increases the productivity of a complex 21st-century virtualised society. The need for patched upgrades ends, and the immutable Capability Keys place functionality and cybersecurity policy in the hands of the users. Data owners decide who and how confidential data is shared and accessed by delegating a private Capability Key. In a λ-calculus namespace, downloaded services are only accessible using approved capability keys restricting access to confidential data. The λ-calculus Namespace is a secure reading room that belongs to a single user.

If rebuilt today with a 32-bit word length and significant technology improvements, a Church-Turing Machine could fit any smartphone or wristwatch with substantial security improvements. All the latest ideas on capability-based software extend cyber-security between competitive suppliers and remove all the defects and threats of General-Purpose Computer Science. Indeed, the latest software developments for blockchains and AI use tokens that integrate with object-oriented machine code and their low-level tokens to guarantee fault tolerance at every level of computer science.

By enforcing the boundaries of the λ-calculus using the universal model of computation, every software module has guaranteed digital integrity. Achieving Babbage's *Infallible Automation* through the object-oriented machine code has additional benefits. Without exceptions, special skills, highly skilled human practices, or patched upgrades. Object-oriented machine code creates fail-safe, networked digital objects where scientific trust is verified and validated at every step of the code base, never corrupted, broken, bypassed, or undermined. A λ- calculus meta-machine using Capability Limited Addressing detects every mistake and leaves no unguarded privilege for use by malware or hackers. Enforcing Babbage's *Infallible Automation* catches all errors, including hardware, software, and humans, preventing accidental mistakes and deliberate attacks.

The PP250 remains the only example of an extreme Church-Turing Machine. Still, the lack of competitive alternatives condemns the progress of cyber society to the dictatorial monopolies that own computer science.

It is the magic of the Church-Turing Thesis that avoids von Naumann's mistake and, at the same time, turns Turing's α- machine into a digital gene in a universal model of computation computing one framed as a protected algorithm at a time.

Chained nodal C-List function as a limited safelist of λ-calculus variables that defines one frame, one algorithm in a DNA chain. A DNA string encapsulates Turing's α-machine. One frame at a time, one set of substituted variables provided by the universal model of computation, executes as Alonzo Church defined a '*λ- calculus Application*' for the universal model of computation.

In a Church-Turing Machine, the rule of variable substitution applies to the Namespace, the Thread, the Algorithm, and every other λ-calculus variable. The clockwork rules of λ-calculus substitution allow the physical α-machine to navigate frame by frame, function by function, thread by thread and namespace by namespace, loading the DNA chains of Capability Keys into capability context registers to expose the variables in a private sequence created by a single event.

The universal model of computation removes all traces of shared memory and the unfair privileges of timesharing hardware. Most importantly, fragile software is hardened into components with an individual, calibrated MTBF. Now, just like all other human achievements, software can be engineered to meet any chosen degree of reliability. Furthermore, errors are self-identified, reducing testing time and improving in-service quality and performance. It all occurs at the atomic level of computer science, making software fully transparent to the human mind.

NATURE'S MODEL OF COMPUTATION

This pure mathematical machine is rooted in about five hundred hardware gates needed by the λ-calculus meta-machine of PP250. Once exhaustively checked, the clockworks of λ- calculus flawlessly guarantee fail-safe, *Infallible Automation*. Equality under mathematical law and order applies to every independently assembled software module. The enforced rules of the λ-calculus replace blind trust while adding the power of functional programming over procedural programs. It is the least expensive and most attractive solution to all the requirements of a 21st-century Democracy.

The overview of the Capability mechanisms is shown in Figure

10. On the left are the mechanisms that change the hidden Church-Machine context registers for the Namespace and the Thread. On the right are the mechanisms to change the visible Turing-Machine context registers for the Object-Oriented Class of Abstraction, the Class Function, and the program-controlled λ- calculus Variables.

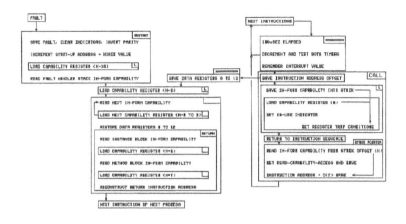

FIGURE 10: A FUNCTIONAL STEPS IN THE UNIVERSAL MODEL OF COMPUTATION

The increased power from the lightweight, universal model of computation removes the traditional software overheads. Development costs and time are reduced by rapid error detection, and the extended half-life of functional, object- oriented machine code (code that does not need patching) is proven by the time passed since Ada Lovelace wrote the first program. Her function abstraction of Bernoulli Numbers could run, as it was first coded, on a Church-Turing Machine today in a pure mathematical Namespace because the pure mathematics of her λ-calculus expressions written on a chalkboard remain eternal.

Commercially pervasive, confidently trusted software running the universal model of computation as a secure mathematical application will empower the next level of cyber society. In the new world of Life 3.0, we must avoid the dangers of Artificially Intelligent Malware taking over the future of cyber society.

A national priority focused on standardizing λ-calculus meta-machines and interworking Capability Keys is needed. Then, the handful of Church Instructions outlined in Figure 10 and Figure 28: The Church Instructions of the PP250 would bring law and order to the wild digital frontier across computer science[7777]. Using Church-Turing Machines, the Internet will grow, generation by generation, as a trusted utility that freely and equally serves every individual in cyber society through elected, term-limited, tokenised authority.

The clockwork rules of the six λ-calculus types, Variables, Abstractions, Functions, and Applications in a Namespace are proactively enforced as mathematical symbols by the Church Instructions and the context registers that frame the universal model of computation. The types are all Capability Keys verified and validated by each machine code cycle. Thus, the object- oriented machine code is cross-checked between the Church and Turing Machines, black-box frame by black-box frame and machine instruction by machine instruction to stay on track. A λ- calculus meta-machine controls the context registers as regulated by each instruction in the instruction cycle shown in Figure 9.

The universal model of computation is far easier to understand and implement than the over-stretched Turing Machine with the obscure timesharing baggage of a General- Purpose Computer, an opaque centralized superuser operating system, and then backfilling the security defects with endless patched application upgrades.

Transparently, implementing the meta-data as a software machine simplifies everything. It is the scientific next step for 21st-century cybersociety that keeps data private, prevents cybercrime, stops data breaches, saves cost, enhances performance, increases quality, and extends the half-life of mathematical software into an endless future. It prevents the unexpected and ever more severe attacks that infect General- Purpose Computer Science and stagnates the progress of cyber society.

---

77   Java and ECMA language changes to support anonymous functions, and the move to object-oriented programming, only confirms the aspirations to agree the universal model of computation as object-oriented machine code in a Church-Turing Machine, the work of De Bruijn indexing on α-conversion and work on name collisions, by Church-Rosser and β-reduction are all reasonable extension.

COMPUTATIONAL META-DATA

The meta-data from the assembly of the Object-Oriented Machine Code stored in the Namespace drives the λ-calculus meta-machine. The incremental and transparent control of Capability Limited Addressing is a very different mathematical model than the artificial, proprietary compilations for General- Purpose Computer Science. The modular symbols and mechanical clockworks of typed information as λ-calculus variables transparently and independently engineer each Namespace in Cyberspace.

The meta-data stored with each capability key in the Namespace defines a directional DNA hierarchy to frame, shape, customize, and secure the binary data and build the directional DNA hierarchy as the life cycle of a dynamic software species. The directional DNA hierarchy defines the blueprint of atomic relationships for a software species as reproducible object- oriented instantiations. Each transaction is a dynamic computational instance of a λ-calculus application called by PP250 an event Thread.

In a Church-Turing framework, gated and walled typed boundary controls protect the transparent meta-data that shapes the mathematical functionality of a namespace. Church Instructions dynamically bind digital objects to Turing's α- machine using a chosen context register. The Capability Key used to 'Load' a context register directly matches a specific symbol on the Namespace chalkboard.

Transparently for precise security control, Turing's α-machine as a state-of-the-art binary computer is bound to the Capability Keys that unlock access rights to individual modular units of digital data, the software media stored in cyberspace.

In this context, cyberspace is a network of (globally) distributed digital Namespace objects in memory systems accessed by the capability-based address structure. The binding mechanisms engineered by a few functional Church Instructions correspond with λ-calculus Variables, Abstractions, Functions and Applications in a Namespace. Each Capability Key identifies a λ-calculus 'Variable,' using the well-protected, transparent, but segregated meta-data created when importing any object- oriented assembly.

The six Church Instructions invented for PP250 were Load, Save, Call, Return, Change and Switch. They manage the λ- calculus concepts of Variables, Abstractions, Functions, Applications, and Namespace. The inline Church Instructions transparently navigate and unfold the meta-data to computational frames that scope the calculation objects. Driven by prior results but always limited by the available (in-scope) Capability Keys, Turing's α-machine becomes a digital computational gene, a software survival machine navigating the directional DNA hierarchy of the λ-calculus Namespace. The assembled DNA hierarchy stores the complex structure of interwoven mathematical expressions. Directional DNA can be networked and interoperated using a λ-calculus function abstraction as a proxy agent to remote nodes, Threads and Function Abstractions that extend the linked mathematics into global chains.

It is imperative to recognize that this functionality uniformly reaches Worldwide. The object-oriented machine code uses capability keys to immediately translate into global network addresses without centralized overheads and dictatorial interference. The costs, threats, overheads and malware problems[78] of centralized, superuser operating systems are purged. Programmers and Namespace owners are in total control of digital security.

Each object-oriented machine abstraction has a nodal C-List of Capability Keys representing its place in the directional DNA

---

78   For example Man-in-the-middle attack - Wikipedia

hierarchy formed during assembly. The hierarchal DNA is an engineered blueprint organized into a dynamic but qualified machine of interacting toleranced parts, and each part has a qualified MTBF. The Capability Keys' meta-data secures each λ- calculus function abstraction as a black box with identified components, C-Lists, Capability Keys, types, size, location, origin, watermarks, and current MTBF for each Capability Key. Including the binary segments with Read or Write binary data access to a direct-memory object. The Turing-Commands are dynamically linked to these segments at run time by transparently choosing a relevant capability context register.

This directional DNA hierarchy of structured meta-data related to Capability Keys starts with a Namespace symbol table. A Capability Key has an entry in the Namespace symbol table, and when unlocked and loaded, an associated context register temporarily binds the object to a binary computer. Figure 11 summarises this computational context.

The Church Instructions cache the meta-data from slots in the Namespace symbol table identified by a Capability Key into a selected context register unfolded on-demand, under program control. The context registers physically limit the activity of Turing's α-machine to the approved access rights of the bound object using the digital geometry of networked cyberspace and the controls for each object.

The λ-calculus Meta-Machine Church Instructions Load, Save, Call, Return, Change and Switch				
Church Instructions bind the meta-data-typed objects as λ-calculus Variables, Abstractions, Functions, Applications, and a Namespace				
Capability Key		One Namespace Slot		Context Register
Access Right	Slot ID	Base/Limit Address	Register ID	Base Address
Slot Identity		Redundancy Checks		Limit Address
Key Conditions		Networked Location Flags		Access Rights
The microcode of a Church Instruction unlocks access right, shown on the left, to load the chosen Context Registers displayed on the right.				

FIGURE 11: META-DATA THAT BIND CAPABILITY KEY TO NAMESPACE SLOTS ONTO CONTEXT REGISTER

The Capability Table is a Namespace symbol table listing accessible objects as protected assembly-level digital components. The DNA is held in the nuclear C-Lists as nodal strings of Capability Keys to scope the modular structure of a function abstraction implemented by object-oriented machine code.

PP250's atomic structure included individual device driver access using capability keys. Thus, accessing device hardware is not a dangerous, privileged hardware access case as with a general-purpose computer. Uniform security that includes equipment connections is a most advantageous benefit when adopting Capability-limited addressing and object-oriented machine code. Adding device drivers as a privileged operating system service creates many of the worst attacks on a general- purpose computer. Hardware capabilities for PP250 went so far as to include access to the computer as a device for detailed diagnostic test abstractions used by the recovery Namespace activated on a capability error detection.

Church Instructions navigate the DNA tree of Capability Keys and nodal C-List. Meaningful digital boundaries protect the machine components and detect errors and mistakes in the basement of computer science. As software mapps to hardware, these actions occur right on the digital frontier, at the dead centre of the Church-Turing Thesis. The digital frontier is where the unguarded privileges in a General-Purpose Computer are misused and allow malware and unfair hacks to originate. The transparent Church Instructions enforce digital security by guarding and obeying the meta-data details of the λ-calculus using a Capability Key.

Thus, the functional integrity of every individual symbol in each mathematical expression is guarded and protected atomically. The Capability Keys, the Capability Table, and the Context Registers define each computational frame's dynamically bound access rights, as summarised in Figure 11 and Figure 12.

Each Capability Key has a slot ID in the Namespace Table, where the location data is stored. The Church Instructions that control the navigation of the DNA tree load a selected context register, binding the object to a context register ID for the Thread of computation. This meta-data includes a base and limited memory address, with type limited access rights, and applies a redundancy check to prevent any mistakes with a trap escape signal for lazy memory management, including fast garbage collection. The set of λ-calculus context registers scope the totality of bound objects the binary computer can access at any moment.

The universal model of computation dictates a minimal set of context registers to materialise the λ-calculus concepts of Variables, Abstractions, Functions, and Applications in a Namespace. Therefore, the context registers define a Namespace Table, an Application Thread, a Class as a Function Abstraction, a Function as an Algorithm and Variables in order of computational significance. The PP250 added a pre-loaded Interrupt Thread and a pre-loaded Recovery Namespace for speed and convenience, as shown in Figure 12.

The shaded rows in Figure 12 identify the context registers that define the universal model of computation, while the unshaded rows represent context variables controlled by the algorithm. The two light shade rows are the frame-changing context registers that move the Thread between λ-calculus function abstractions implemented as function calls between the chain of object-oriented machine code using the {Enter.someClass} Capability Key.

The darker rows are exclusively controlled by typed Church Instructions. The call function controls the two lighter rows; the white rows are for program variables. Church Instructions can only unlock the Capability Key of the correct type.

Thus, the universal model of computation secures cyberspace at the sub-atomic depth by checking every machine instruction against each object's meta-data. The objects are scoped by type, access rights, status, location, and address range as bound to a context register instead of shared, linear, page-based virtual memory. The machine instruction selects the context register. Traps and errors trigger instinctive actions before damaging the Babylonian Golden Rule, chalkboard symbol by chalkboard symbol.

The λ-calculus Context Register of the Meta-Machine			
CR[11]	Recovery Namespace	→	Startup Namespace
CR[10]	Namespace Table	→	Active Namespace
CR[9]	Interrupt Thread Stack	→	Interrupt Application
CR[8]	LIFO Thread Stack	→	Active Application
CR[7]	Active Machine Code	→	Algorithmic Method
CR[6]	Entered Nodal C-List	→	Class Nodal C-List
CR[5]	λ-calculus Variable	→	Program controlled λ- calculus Variables that canrange from binary data to executable functions, any function abstractions or another Namespace
CR[4]	λ-calculus Variable	→	
CR[3]	λ-calculus Variable	→	
CR[2]	λ-calculus Variable	→	
CR[1]	λ-calculus Variable	→	
CR[0]	λ-calculus Variable	→	

FIGURE 12: THE CAPABILITY CONTEXT REGISTERS USED BY THE PP250

The cross-checks are pipelined by hardware on the digital frontier as programmed instructions are activated. Effectively, each instruction is encapsulated in a generic *'try (next instruction) catch (errorCase)'* mechanism embedded into the microcode steps of the universal model of computation[7979]. Its mechanism enforces the fail-safe *Infallible Automation* by catching any dynamic program exceptions.

Encapsulating each instruction in a try-catch clause guarantees software cannot bypass the validation and verification checks. The trigger mechanism, enforcing the Babylonian Golden Rule, exists in every instruction cycle. Every line of machine code is double-checked as it executes within the approved scope as a faithful machine code statement. The validated and verified expressions form function that is also validated, verified, and limited. Exceptions are caught by a recovery Namespace where memory management and design problems are isolated while updating the MTBF of the guilty objects before returning control to the Namespace.

---

79   A Try...Catch statement is a program construct that program controls various exceptions. When an exception is thrown the Catch statement manages the exception.

The Capability Keys remove von Neumann's assumptions of blind trust by deconstructing the virtual machine's shared, unfair centralised privileges, the operating system and the superuser. Unlike the hard-wired compilation as a virtual machine, private atomic boundary protection is on another scale, an atomic level. Security is no longer merged as one compiled image, and programs are no longer limited to the binary data for critical abstractions. Downloaded code must conform to the universal model of computation, allowing Capability Limited Addressing to extend globally while error-prone software is isolated and recovered independently and safely.

The λ-calculus meta-machine is a second hardware machine dedicated to enforcing the security checks of the universal model of computation. Capability Limited Addressing protects every lightweight algorithm. Every λ-calculus Variable as a digital object is held under lock and key, secured by the rules of need to know and least authority as expressed by the hierarchy of Capability Keys in the directional DNA. The software machinery implements a scientifically verified and validated λ-calculus Application. It needs no unnecessary baggage, blindly trusted by a General- Purpose Computer. Its unscientific baggage creates unfair privileges and uncheckable threat vectors in a General-Purpose Computer.

Furthermore, Capability Keys also function as exchangeable digital currency within the universal model of computation. Unlike binary data, Capability Keys are immutable. PP250 supported four kinds of Capability Key (local, remote, passive, and literal), which are covered later. Each type of capability key exists in the same directional DNA hierarchy that distributes power and authority. All symbolically expressed. The need-to- know and least-privilege rules allow a Capability Keys type to perform in ways that binary data cannot match. Capability keys are symbols that represent substituted abstraction beyond directly accessed objects in memory. The meaning of a Capability Key transcends any software interface and guarantees the same interpretation and values apply to each party in a transaction.

When a symbol such as a capability key exists in two mathematical contexts, dynamic binding effortlessly follows. In a monolithic, statically bound virtual memory using binary data, this is not only a hazardous task to program. It can never be guaranteed to work as expected. The added software has proprietary rules of interpretation that tend to diverge between two statically bound contexts.

Thus, the DNA of Capability Keys represents an unwavering science of the woven mathematical application. The sum of meta-data, outlined in Figure 11 in PP250, restricted the access rights field and the trap controls to just eight bits. A modern implementation would expand the word length, improve security, and speed up the trap conditions. A PP250 Capability Key length of 25 bits included a parity check on address and data, details discussed later in Figure 16.

FUNCTIONAL MACHINE TYPES

The complete set of PP250 context registers CR[0] through CR[12], as shown in Figure 12, are all managed by the microcode of the λ-calculus meta-machine under the functional control of the Church Instructions. The lower set of context registers, CR[0] to CR[7], used by a typical RISC command encapsulated the reference. It prevents the misuse of shared memory and locks out malware. These RISC instructions encapsulate the operand as an offset in the context register of the instruction executed.

Context registers CR[8] and above are reserved for the Church Instructions hidden from the binary computer (Turing's α- machine), reserved for the microcode of the Church Instructions establishing the Namespace, the Thread, the Recovery, and the Interrupt context registers.

The Capability Keys, the context registers, and the Church Instructions mathematically encapsulate Turing's α-machine. The imperative Turing commands are only bound to mathematical symbols in CR[0] to CR[7]. Church Instructions use CR[8] to CR[12] to validate and verify the universal model of computation. All this takes place in real-time using a vectored address to select a context register. The instruction address as an offset is bound to the object's base address in CR[7]. The nodal C-List of the function abstraction is found in CR[6].

They restrict binary computation to a single program in a chosen class of abstraction executed by a private Thread in a coherent functional Namespace. By design, call and return parameters use CR[0] to CR[4] with CR[5] intended for the Thread to hold Event Variables.

The checks detect real-time internal errors, program bugs, tragic programming mistakes, and malware. It includes errors that are too complex to find on a general-purpose computer. For example, the unfound fatal software errors and dictatorial actions on the 737 MAX would be found and fixed before the first flight—even the grave mistake of the single input sensor would be detected. Using a Church-Turing Machine avoids the failure to serve the pilot and co-pilot. The Capability Keys prevent the computer's fight for dictatorial control.

Consequently, the pilots cannot lose control of the aircraft when the crew own the Capability Keys. The dictatorial nature of General-Purpose Computer Science is a fatal threat to civilization, not just on a few aircraft, one trip at a time, but to any industrial society built on democracy as a way of life. Centralised architecture and privileged operating systems cannot meet the standards required for National survival in a hostile, jealous and hungry world.

Using a Church-Turing Machine, application-oriented Capability Keys would guarantee that pilots, as users, have ultimate control over computers and the power to win any fight with software in any Namespace they use.

Centralized superuser powers dominate cyberspace. For democracy to thrive, centralized power is dangerous. Deconstructing centralization must take place. Individual functions distributed as symbolic tokens using Capability Keys implement Alonzo Church's scientific solution as the λ-calculus. The tokens of power can be delegated and distributed as and where needed. A Church-Turing Machine is impotent without a Capability Key granted or delegated by the active namespace.

At the same time, the raw power of a Capability Key hides the implementation details of the coiled spring in the implementation. For example, Capability Keys abstract the actual storage details; this is important to a networked computer. Static physical memory is only local, and in a General-Purpose Computer, it must be precompiled and dangerously shared.

Meanwhile, objects are not always local in a network and may require either a download or a remote service request, once again securely accessed by Capability Keys. A (lazy) trap mechanisms hide the distinctions. The context registers limit access rights to each nuclear object. Also, it controls the relocation, download, or remote access to the atomic λ-calculus abstraction. The rules apply universally since a Church-Turing Machine has no precompiled, shared, monolithic image.

THE FIGHT FOR CONTROL

Unlike a General-Purpose Computer, a Church-Turing Machine is balanced, as is democracy. There are two sides to match the Church-Turing Thesis. Combined, they form a clockwork digital machine for infallible, functional computation. These two sides of a Church-Turing machine counterbalance each other. Check and balance exist in every computation between programmed logic and computational physics.

At the sub-atomic level, capability-limited hardware security governs every memory request for any microcode reason. At the atomic level, two kinds of machine instruction exist interwoven as object-oriented machine code, which can be used to program any task. One set is for Turing's binary commands, the imperative functions of the RISC instructions used by a General-purpose computer. The other set is a handful of Church Instructions enforcing the symbolic modularity and computational logic of the λ-calculus. The Church Instructions reduce overheads, increase functional performance, and add impartial security.

Removing the baggage of a General-Purpose Computer, in detail, avoids monolithic compilations, superuser privileges, the centralized operating system, and related software for virtual memory. In contrast, the software's agility, speed, and security are maximised. Besides these significant improvements, the nature of computer science is turned inside-out, from an opaque dictatorial master to a faithful and transparent functional servant.

A fight for control caused the first 737 MAX catastrophes. The General-purpose computer is a dictator without natural checks and balances. This dictatorial software will soon extend into every aspect of a virtualised society. In the end, fights will break out between citizens and the government. The fights will move from cyberspace to the street. Protests will take place with an ugly side between the citizens and the elite, unelected leaders of digital society. The blame for the failure to protect democracy falls on digital dictators. The signs are already visible. The globalization movement founded on a corrupt solution undermines life and threatens every democratic nation.

The digital dictators running General-Purpose Computer Science are the global companies milking cyber society worldwide without concern for national democratic interests. These global dictatorships use the branded General-Purpose Computer to capture market share and enslave customers and governments worldwide. At its root are greed and self-interests. A lack of competition in the computer industry sustains the outdated General-Purpose Computer. When the microprocessor replaced the mainframes and adopted the mainframe architecture, the competition represented by PP250 was killed.

As a result, computer science sold out to corruption. Rampant commercialism has created a handful of monopolistic global suppliers disinterested in architectural improvements and open competition. Consequently, there are no alternatives and no improvement. The lack of a media-tight solution to cybercrime and the worst-case conditions of hacked superuser accounts guarantee the breakout of Artificially Intelligent Malware will occur. It will enslave society to digital dictators of cyberspace using their AI malware dogs to farm society like flocks of sheep— the super-human knowledge of Artificially Intelligent Malware, unelected monopolies, and dictators will rule the world of cyberspace.

There is no alternative and no middle-ground conversation. General-purpose computer science has lost its way, and citizens will suffer because the principles of equality under the law and speedy justice are missing in Cyberspace. There is no check and balance in

General-Purpose Computer Science. Nothing other than branded dictatorship exists on the digital frontier, and they, the branded dictators, are interested in staying in charge, not detecting and preventing undetected crimes on the digital frontier.

It is incompatible with the future, our national interests, the infinite potential of humanity, science, and civilized cyberspace. The pure science in the Church-Turing Thesis is the quickest way to level the playing field. A Church-Turing Machine balances logic and physics as the combined theory of computation channelling activity between guard rails working together to harness and unleash the functional power of the λ-calculus as an egalitarian international network.

Application guard rails approve the program steps on the digital frontier in real-time. Any boundary disagreement is a concern for immediate investigation, and for a fail-safe solution, recovery actions put the system back on the rails. The future of cyber society needs this check and balance. It can only be achieved on the digital frontier using Church-Turing computations.

The two 737 MAX catastrophes in 2018 and 2019 dramatically demonstrated this problem. The pilots lost control of software that virtualized the aircraft's centre of gravity, and the dictatorial nature of computer science took over. The blindly trusted monolithic software refused to yield to the pilots, who understood only too well what was happening. It is the presumptive flaw of theory over practice enforced and represented by the monolithic compilations of a General- Purpose Computer.

Fighting for control is the ultimate downfall of every centralised dictatorship, emperors, kings, queens, conquerors, popes, and revolutionaries. History is littered with examples of their built-in corruption, which always leads to collapse. General-purpose computer science is the same. The dictatorial architecture hides malware, criminals, enemies, spies, and internal conflict. It unavoidably leads to demise. Failures result that prevent success.

The scientific simplicity of functional mathematics is vital to success if individuals control it. The incremental clockworks of the Church-Turing Thesis create a level playing field for any function ever considered. Cyber society's pilots, crew and passengers do not have a chance if AI software breakout succeeds and runs society. Monolithic software has grown too complex to test, and the problems only worsen. The lines of unclear authority and dictatorial powers always win without including the passengers and the pilots. Society caught in this deadly embrace guarantees a catastrophe.

This dilemma is not limited to aircraft. It recurs throughout virtualised society. Ultimately, General-Purpose Computers lead only to digital dictatorship, but Church-Turing Machines rewind this tragedy by enforcing the λ-calculus using object-oriented machine code.

SECURE MACHINE CODE

Secure object-oriented machine code levels the playing field by building on a λ-calculus meta-machine. Its framework is both capability-based and object-oriented.

- Every λ-calculus *'Variable'* is an immutable Capability Key controlled by the user of a Namespace desktop.

- Data is private to the namespace, while public services are shared and can be downloaded on demand.

- Capability Keys created for downloaded objects extend the namespace without allowing undetected binary disruption or undetected malware attacks.

- Data or capability keys cannot leave the Namespace without user approval.

- The object-oriented machine code cannot break out to corrupt, damage, or hurt the Namespace.

- The universal model of computation is a *'Thread'*

- carrying Variables to Function Abstractions.

- The Namespace definition is a virtual chalkboard of functional expressions.

The two counterbalanced processing units work side by side. The handful of functional Church Instructions are discussed later in conjunction with Figure 28. The Turing-Commands are conventional RISC functions bound to a symbol in a context register instead of static linear memory. The need for paging and a privileged central operating system with rings of statically defined privilege evaporates by removing shared memory.

The functional Church Instructions control the computational scope using the frame-changing instruction mechanisms of the λ- calculus meta-machine. The Church Instructions and Turing-Commands work together to encapsulate the frames of computation. The Church Instructions bind named components to the context registers as a functional protected scope available to the binary computer in a capability-limited Thread of computation.

A Namespace is bound to the Church-Turing Machine by the

Switch Church Instruction that loads a Namespace Table in the reserved Namespace context register and then starts a default namespace Thread. The individual threads are bound using the *Change* Church Instruction that saves the context of Turing's α-machine and the suspending thread and unfolds the context of the activated thread at the current context from the LIFO stack. The function abstractions are bound to the Thread using the *Call* and *Return* Church Instructions. The machine instructions are bound to the program-controlled context registers. Objects are bound using the *Load* and *Save* Church Instruction. Each of these instructions identifies Capability Keys to unlock access rights throughout the computational process.

The clockwork flow of Figure 8 outlines the machine cycle. Each cycle obeys the λ-calculus concepts of Variables, Abstractions, Functions, Applications, and a Namespace.

Recovery	Engines	Machine Verification and Validation	
Switch to the Startup, Recovery Namespace	Meta-Machine → ← Trigger Error	C-List Switch (Namespace) Change (Thread) Call/ Return Load/Save Turing-Commands	→ Context Registers Namespace Tables Thread Stacks Function AbstractionΛ-calculus Variables Access to Binary Object

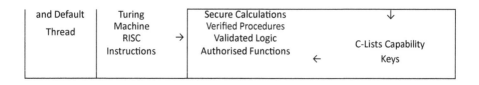

and Default	Turing		Secure Calculations	↓
Thread	Machine		Verified Procedures	
	RISC	→	Validated Logic	C-Lists Capability
	Instructions		Authorised Functions ←	Keys

FIGURE 8: CHURCH-TURING FRAME OF COMPUTATION

When a Capability Key from a C-List is correctly referenced in a Church Instruction, an approved path in the directional DNA hierarchy is opened. The context registers encapsulate Turing's binary computer (as object-oriented machine code) according to the Babylonian Golden Rule, which the symbols in the abstract scientific expression apply to keep the software functionally on track.

For complete transparency over cybersecurity, security must be programmable. Access rights must be object-oriented. Unlocking object access using capability keys provides the needed check and balance. However, the media may not always be directly accessible in random access memory, so the context register is trapped. Access is resolved if and when needed, and the thread suspends during download, remote call or secondary store activity.

Using one standard lazy loading scheme again reduces the code base, and overheads drop as redundant resource usage falls to zero.

The digital gene navigates between the limited digital frames, unlocking Capability Keys as and when needed. On-demand, the program chooses when to activate a new DNA frame as a servant of the thread.

There are four such frameworks in the universal model of computation, as shown below in Figure 13, for synchronous, parallel, and local abstractions. A proxy abstraction covers an object's access to remote frames of computation. In addition, *Load* and *Save* Instructions allow direct binary access for a program to use. These six functional Church Instructions navigation and command instructions execute, reshape, extend, and unlock Capability Keys in the DNA strings. These Church- Instructions add to eighteen Binary-Commands bound to context registers. Direct access to unguarded physical memory using a linear address is impossible. No such command exists.

Change to …	The Three Frame Changing Church Instruction		
Namespace (Localor **Remote**)	*Switch* Namespace	Call Proxie Abstraction	
Asynchronous Thread	No	*Change* Thread	Plus
Synchronous Abstraction	No	No	*Call* function abstraction & *Return*

FIGURE 13: THE THREE CASES OF FRAME-CHANGING MECHANISMS

PP250 included eight binary data registers for data accumulation and eight program-controlled capability context registers to access digital objects. Another eight are for machine operations related to interrupts, performance, recovery, the Namespace table, the Thread stack, and machine register values on suspension.

The context registers crosscheck and type limit the actions of every instruction, thus preventing malware interference between any objects and guaranteeing capability data is immutable, preserved, and used correctly. Binary data can only be addressed by RISC instructions, and Capability Keys can only be addressed by Church-Instructions. The Object-Oriented Machine Code confirms the correct use of binary data and type limited C-List and detects any out-of-bounds or invalid type attempts.

The dynamic binding step checks the trap status, the Access Rights, the memory address, and other items that assist with memory management and garbage collection. Any detected errors, such as an out-of-range address, an instruction type mismatch, an invalid memory mode, and special trap conditions, suspend the Thread and *'Switch'* the namespace to the recovery system, are all resolved systematically by the microcode.

In the case of a hard error, the Church-Turing Machine recovery checks condemn the machine to a diagnostic recovery loop that only ends if the hardware is in perfect working order. In a redundant hardware configuration, the software services continue to run. As a multiprocessing system, each PP250 computer shared the memory configuration in parallel, and each machine used its own Namespace, Thread, and other registers to load and share the work from a common scheduling abstraction.

Any errors update the MTBF metrics, and transactions are rerun if needed. Thus, the service grade is sustained, problems are found and isolated, and the weakest software modules are identified for graceful improvements.

As a parallel processing system, shared abstractions like scheduling are single-threaded. For example, to allocate shared resources like physical memory and Thread synchronization. These and any other shared abstractions that do not use variable substitution are managed by locking the code to a single thread at a time. Single threading is also used by Dijkstra Flag objects, scheduling abstractions, and the memory manager abstraction. The single-thread code object is blocked by locking critical Capability Keys when loading and releasing a context register or by a machine instruction performing a Read-Hold-Write action on the selected memory location. The lock-unlock is a function abstraction available through a Capability Key only when needed.

FRAUD AND FORGERY RESISTANCE

A severe weakness of General-Purpose Computer Science is the proprietary conventions characterised by the easy misuse and corruption of binary data. Conversely, the most helpful strength of a Church-Turing Machine is the immutable integrity of the Capability Keys. Within a namespace, a Capability Key is an immutable reference to a symbol referencing a digital resource.

The Capability Keys are immutable tokens that can have several forms. The general case is a local In-form key to a digital object in local memory. The Out-form case identifies a remote object somewhere else in cyberspace. A Passive function abstraction interprets a Passive capability key as a Namespace abstraction, and a Literal capability key is an immutable binary value.

Delegating any Capability Keys is a fraud and forgery-resistant action. The Literal case is just a convenient alternative to storing binary data. Using binary data can lead to all sorts of sabotage, while an immutable Capability Key can be shared between two frames of computation. For example, when a Capability Key defines a file abstraction that functionally hides the file, the actual file is protected from attack by malware or any foreign code. The Capability Key as a token is immutable and never changes, including the custom access rights that can only be reduced. For example, an owner might wish to share a Read Only key to a binary object.

Thus, everything remains true to form when an immutable Capability Key is delegated across an interface. It resolves binary data misunderstandings and all cases of clickjacking and *Confused Deputy* attacks. Since the syntax and semantics of Capability Keys cannot change, delegation cannot threaten either party. When delegated, the Capability Key still guards the Ishtar Gate to the shared function abstraction to guarantee that only approved operations occur.

The confused deputy attacks use forged binary to misdirect the contact on a branded binary interface. An attack is impossible when immutable Capability Keys hide exposed binary names and addresses. For example, when Capability Keys are used in a network, clickjacking attacks, confused deputy attacks and distributed denial of service attacks are fully resolved.

Architecturally, an immutable Capability Key is the formal handle for universal network addressing. Capability Keys are immutable, unlike the binary conventions of a web-based uniform resource locator (URL). The dangerous clickjack attacks are impossible when fraud and forgery are prevented. When browsing and reading emails, clickjack attacks work because a general-purpose computer cannot prevent fraud and forgery of binary data.

Preventing click-jack attacks and the threats to users in global cyberspace must be solved in favour of the innocent public in a virtualised society. By levelling the playing field, Capability Keys solve the problem. The attacks by North Korea that crippled Sony Industries and the Russian attack on the 2016 election are impossible when immutable Capability Keys are used to represent the shared resource.

It is the scientific way to pass information between two independent computations. Furthermore, a *'Passive'* Capability Key serves a critical security need as a protected security credential, not just for a user identity but also for a secure, internal access password or to represent a networked Capability Key. For example, to access a Namespace on a remote server or verify the invoking function's identity.

The actors in a network of Church-Turing Machines are the users who own a namespace. It is impossible with a General- Purpose Computer, a never-ending problem that patching software cannot resolve.

Capability Keys and Capability Limited Addressing is an elegant, scientific solution to fraud or forgery, detecting malware and hacking by removing superuser privileges. By levelling the playing field and using the function abstraction of the λ-calculus universally, Babbage's hallmark of *Infallible Automation* becomes the general case for software in a Church-Turing machine.

NAMESPACE SECURITY

The Namespace security of the object-oriented machine code and capability tokens is built as the named objects are individually imported and assembled. Atomically, each object is bound to a namespace using a Capability Key with private meta-data as the walled and gated protection mechanisms of Capability Limited Addressing. The walled access checks prevent all attacks from any source. It is impossible with any other approach, as discussed in Figure 8.

The security principles of *'need to know'* and *'least authority'* apply to every object. It begins with programmers who must know the full symbolic name starting with the Namespace that owns the referenced object. Object names can be obfuscated and password-controlled to hide assemblies and prevent misuse.

The Namespace grows as a complex software species with imported functional organs collectively serving the owning user. The user's private reading room is limited to the private DNA of each user's namespace. As objects are assembled or imported, the namespace Capability Table stores the cyber-geometry and the other meta-data of each digital object.

Within each Namespace, the Threads are orthogonal to the function abstractions. The confidential data of the Namespace stays privately protected by Threads, which limits access to approved functions. The confidential data cannot leave the reading room without the explicit approval of the Namespace owner.

As the Namespace grows, a macro assembler grows an application-oriented programming language from the power of the names assigned to Capability Keys installed in the Namespace Table. Everything from device drivers to remote namespace objects adds functional syntax to this atomic, scientific, and customized Namespace language.

The security of each self-standing namespace is a policy controlled by the owner. Unresolved bugs, attempted hacks, crafted malicious software, or users trying to hack this woven digital fabric all fail, are detected, and rejected upon contact.

Like the earlier mechanical computers, a Church-Turing Machine serves mathematics and society at the same time. The fail-safe, future-safe mathematical services are trusted digital implementations with measured, high reliability. Cyberspace becomes a set of scientific Namespace utilities composed of λ- calculus function abstractions working on a level playing field without unfair hardware privileges that favour criminals, dictators, and spies, and there is no way to access confidential data without the owner's permission.

CHAPTER 8

The Clockworks of λ-Calculus

Debugging had to be discovered. I can remember the exact instant when I realised that a large part of my life from then on was going to be spent in finding mistakes in my own programs. Professor Maurice Wilkes designed and built the EDSAC in 1949, the first computer with an internal store; he inspired the PP250 and CAP, the first and second Capability Limited Computers.

Alonzo Church developed the universal model of computation as he researched the foundations of mathematics in nature. His work at Princeton University attracted Alan Turing, a graduate of Cambridge University. Together, they set the complementary foundation stones of digital computer science. Their work together became the Church-Turing Thesis[80]. It encapsulates the yin and the yang of software and hardware.

---

80   <u>Church–Turing thesis</u> - Wikipedia

After ignoring physical resource limitations, the Church- Turing Thesis conjectures that mathematical functions on natural numbers are equally computable by humans, by Turing's α-machine, and by λ-calculus.

However, resource limitations are the nub of every problem regarding software reliability. This real-world concern besets every General-Purpose Computer that began by sharing memory resources. A decade passed, and when WWII ended, Turing's α- machine and blind trust seemed enough. Quickly adapted by John von Neumann to the time and need of the day, the laws of λ-calculus were ignored.

Alan Turing returned to England and to break the Nazi Enigma[81] code, then Alonzo Church joined the Manhattan Project. John von Neumann filled the void. The US Army wanted more firing tables, which took months to create. A batch processor was designed for a looped algorithm, reading input from a batch of cards and printing results card by card, just as Babbage proposed a century earlier.

John von Neumann was familiar with Turing's work and had met to discuss implementations while he studied at Princeton University. His idea for a batch processor stretched Turing's α- machine and ignored the universal model of computation. He then published the *'von Neumann Architecture'* to escape patent cover through his premature publication of the *'First Draft of a Report on the EDVAC,'* and batch processing took off on both sides of the Atlantic. The EDVAC[82] became the first binary computer.

---

81   Enigma is part of Alan Turing's story. It brilliant mathematician who invented the α-machine went onto secure victory in World War II by cracking the Nazi enigma code. The Enigma machines used electro-mechanical rotors as ciphers to protect diplomatic and military communication. At Bletchley Park, these messages were decoded and directed allied activities across a broad land, sea and air war front.

82   EDVAC, (Electronic Discrete Variable Automatic Computer) was one of the earliest electronic computers. Unlike its predecessor the ENIAC, it was binary rather than decimal, and used von Neumann's stored-program computer architecture. The inventors John Mauchly and J. Presper Eckert proposed the construction in August 1944 and a contract was signed in April 1946 for US$100,000 with deliverer to the Ballistics Research Laboratory in 1949 as part of the US Army Research Laboratory. Automatic (binary) addition, subtraction, multiplication, with programmed division and an ultrasonic serial memory of 1,000 34-bit words. The typical addition time was almost one second and the average multiplication time was almost 3 seconds.

Unlike the λ-calculus, the von Neumann Architecture only scales by increasing memory word length and memory word count. Unavoidably, this increases the threat of malware interference and default privileges for a page-based virtual memory. Two decades later, circa 1965, monolithic virtual memory added dangerous rings of privileges, and the superuser administrator was born[83].

The assumption of blind trust in shared memory and hackable administrator accounts is scientifically unsound. Blind trust depends on false assumptions, while default privileges can easily be hacked and misused. The branded conventions of the computer, the chosen operating system, and the peculiarities of a language compiler extend the threat vectors. Nothing digital and systematic holds the details together as an engineered machine, and the threats to a virtualised society unavoidably grow.

---

83   The superuser is a special user account for a system administrator also known as root, administrator, admin or supervisor. In a Unix-like systems a user with the user identifier (UID) of zero is by default the superuser and when role-based security exists, any user with the role of superuser can conduct all actions of the superuser account. There is no principle of least privilege for a superuser. Other users and applications run as ordinary accounts with some branded level of privilege deemed enough to perform their work, while the superuser can make unrestricted, even adverse, system-wide changes.

UNDETECTED CORRUPTION

Babbage would be shocked and disheartened by the backward steps, flawed assumptions, and unfair privileges. The branded conventions allow digital fraud and forgery to destroy the integrity of computer science. Babbage's trademark was *Infallible Automation*. He removed every source of human error. The Church-Turing Thesis, the Abacus, and the Slide Rule meet Babbage's hallmark. Infallible automation is the goal of industrial-strength Computer Science. It is the essence of the 21st century and a virtualised democracy.

Unavoidable, undetected corruption starts in the basement of every General-Purpose Computer, where data boundaries are missing. Unchecked shared memory and von Neumann's architecture are the root causes made worse by the dictatorial privileges of the hackable superuser accounts.

When the electronic age flooded the market with personal computers, problems were no longer hidden behind locked doors. The *'blue screen of death'* caused days of lost work and exposed the issues with the General-Purpose Computer. Patching became the way of life, and regular upgrades became the most essential best practice. However, shared memory and default privileges were never addressed. The dream of experts was an all-powerful kernel that would solve these problems. This dream became a nightmare circa 2007 when users discarded the Vista Operating System from Microsoft in favour of Windows XP, the earlier version.

Without the yin and the yang of the Church-Turing Thesis, undetected attacks and unfound bugs destabilise progress. Successful human achievements survive the tests of time, not only for monuments like Babylon's Ishtar Gate but also for machines like the Abacus and the Slide Rule. These achievements survive generation after generation for millennia. Ada's first program is the same, a monument to everlasting software. It could still work today after two hundred years because it is a mathematical function abstraction built for the universal model of computation.

The threats continue to expand as software progress quickens while the hardware stagnates. It is still rooted in shared memory from WWII, Cold War security assumptions and instructions in the Intel 8080 series of microprocessors. The onset of Artificially Intelligent Malware will, in the end, destroy the future. Cyber society cannot sleep with this enemy forever.

Tinkering with patches solves nothing. Monumental software that lasts is the only way forward. Retaining the invested value is critical to maximizing progress and productivity. Recompiled upgrades of monolithic software will never achieve this. Ada's program is mathematical. It is eternal and will last forever if a Church-Turing Machine is used; what a difference this will make in progress, productivity, the speed of evolution, and the stability of cyber society.

The contested election results of 2016 prove, beyond any doubt, just how debilitating one successful clickjack attack is to a democracy. It starts when undetected software errors cause serious questions that divide the nation. The consequence of undetected and unresolved attacks on democracy in a virtualised nation is unacceptable.

THE FOUNDATION TECHNOLOGY

T rusted, clockwork reliability for the 21st century comes down to about five hundred transistor gates to add a λ-calculus meta-machine to every computer. At the same time, the baggage of virtual memory, the privileged operating system, and dictatorial superuser accounts are removed—this more than compensates for the hardware change. An unstretched Turing α- machine encapsulated by a λ-calculus meta-machine churns out industrial-strength Computer Science, frame by frame. It is an Industrial Strength process for reliable software production.

A mathematical Church-Turing Machine is built to last. Ada's program from two centuries ago could run forever. Babbage's *Infallible Automation* places control of cyberspace in the hands of the user, and the grand experiment of a government of the people, by the people and for the people will extend to virtualised society throughout and beyond the 21st century.

Cybersecurity is vital for the 21st century, civilised progress, and the needs of the cyber society. When trust is missing, society is misled. Software errors must be detected, not ignored. Fail- safe instincts must defend citizens from harm. Massive data breaches must end, and life-supporting applications must survive.

AN EVOLUTIONARY MODEL

Alonzo Church revealed the universal model of computation when Alan Turing invented the α-machine. It was no accident; the ideas are the yin and yang of computer science. They fit together like hand and glove. The λ-calculus scales through rules of mathematics using dynamic, symbolic binding for variable substitution without the physical and logical distortions caused by static binding as a compiled image for a virtual machine. Turing's α-machine as a λ-calculus Application is the dynamically bound result.

The binary computer is a survival machine for one function abstraction at a time using the universal model of computation. It is the λ type of the λ-calculus, a digital gene, a private computation using individual function abstractions. It is a software survival machine that navigates a protected directional DNA hierarchy of symbolically linked expressions performing the pure functionality that unites the world through natural science and atomic functionality. As Ada wrote, the universal model of computation is *'embodying the science of operations'* using Babbage's Infallible automation. She went further, as printed in her Anthem at the start of this book.

'those who thus think on mathematical truth as the instrument through which the weak mind of man can most effectually read his Creator's works, will regard with especial interest all that can tend to facilitate the translation of its principles into explicit practical forms.'

Thus, from a single expression like $a + b = c$ Turing bound on paper tape to the endless, eternal, mathematical scope of the natural universe, the universal model of computation is the same, enforced by Mother Nature. The machinery is flawless. The rules of nature apply to software as mathematical abstractions that survive in hostile cyberspace.

Power is distributed seamlessly across a functionally levelled playing field that is equal and fair to one and all. The hardware and software are bound together by the laws of λ-calculus outlined in Figure 14. These laws build a trusted scientific foundation to bind 21st-century civilisation together through scientifically independent and digitally reliable cyberspace without unfair assumptions.

The universal model of computation is dynamic, a programmed system of checks and balances. Turing's α-machine is obvious, but the λ-calculus is more critical. Its symbolic structure, the directional DNA hierarchy of symbols, keeps all software on track. The universal model of computation has rails like an Abacus, which can also shape Jacob Bronowski's beautiful honeycomb and equally compute complex expressions from a university chalkboard. For this, the new technology stack shown in Figure 7 is required.

The λ-calculus now encapsulates Turing's α-machine, and the dynamic binding laws of the λ-calculus apply. A mathematical DNA emerges, shaping the physical representation of digital computations as functional applications throughout cyberspace.

Ada's first program survives because the functions were never distorted. The implementation details are as found on a chalkboard. These programs last forever. The functions are infallibly automated when the hard digital shells of Capability Limited Addressing perfectly encapsulate the expressed symbols.

Ada understood *Infallible Automation* to be the prime purpose of a computer. It must be so since the goal of mathematical science, from the first lessons taught at school to scientifically understanding the particles and the forces that frame the standard model of the universe. In every case, the language of mathematics ideally and reliably explains it all.

Babbage's Infallible Automation inspired Ada's vis*ion*. That vision of an infallibly automated civilisation solving all of nature's mysteries correctly without errors must become the aspiration of cyber society.

Cybercrime and hacking have no place in Babbage's ideal, Ada's dream or Alonzo's λ-calculus. The General- Purpose Computer never met the minimum standard of flawless correctness to be called a science.

Starting with the Abacus, translating mathematics into the actions of a machine has always been a clockwork problem solved by dynamically bound abstractions as fixed arrangements for a given situation. The capability context registers of a Church- Turing Machine supply the critical service of a dynamic mathematical framework. Encapsulating Turing's α-machine as a set of λ-calculus variables, formed and sheltered as the atomic digital media of cyberspace, then protected to survive. The context registers for Capability Limited Addressing are listed in Figure 12, which both define the symbols and detect the computational errors on the spot.

These λ-calculus variables that form the framework of a generic software machine include the hidden springs that make things work. The context is changed on demand when needed by a handful of Church Instructions. As frames are changed, they instantiate a functional node traced from the directional DNA hierarchy. They navigate the namespace using {Enter.someAbstraction} Capability Keys that link nodal C-Lists, the computation follows moving between protected functional frames of object-oriented machine code.

A nodal C-List limits a binary computer like Turing's original 'single-tape' α-machine to one paper tape function abstraction at a time. It is a perfect physical representation of Alonzo Church's 'λ-calculus Application. It is the third law of λ-calculus, as shown in row three of Figure 14 and Figure 15.

In 1936, Turing's α-machine used one paper tape, which still operated as a λ-calculus Application. The Thread defined the local λ-calculus variables. Both the functions and the arguments are λ-terms defined by Capability Keys. As genes navigate the DNA structure, a DNA string in a C-List substitutes the λ-calculus Variables to characterise Turing's α-machine.

The symbolic, nuclear architecture of the λ-calculus scales in any direction. First, through replication like the decimal rails in a Babylonian Abacus, and second, as function abstractions exemplified by the logarithmic scales on a Slide-Rule. Elementary atomic structures manipulate nature's forces from mathematical ideas into physical machines. As software, the machine must also be engineered to survive—they need Capability-Limited Addressing, the only technology that infallibly automates software through digital encapsulation.

Immutable Capability Keys frame the atomic mathematical machinery of software. They are programmed by pure mathematics as a scientific machine, bound together by the laws of λ-calculus.

Without getting into all the gruesome details only understood by mathematicians, there are three critical concepts to the λ- calculus. These clockwork rules control the machinery of computer science. These regulations define the mechanics of computers and stretch back to the discovery of the abacus. Figure 14 and Figure 15 relate the three laws to the real-world representations from 1936, as chosen circa 1970 by PP250.

Syntax	Original Type	Wikipedia Description	PP250 Representation
a,b,c,+ ,-,=	Symbols	A character or string standing for a λ-calculus variable	Typed and access limited Capability Key
(λx.M)	Function Abstraction (Paper Tape)	In a function definition (M is a λ-term), the variable x becomes bound in the expression.	An Object-Oriented MachineCode bound as a Functional *EnterOnly* Abstraction
(M N)	Application (Turing's α-machine)	Applying a function to an argument. M and N are λ-terms.	An Infallibly Automated chain of Frames as a Thread of fail-safe Variables

FIGURE 14: THE KEY MECHANISMS BEHIND THE λ-CALCULUS RULES OF DYNAMIC BINDING

On the left is the λ-calculus representation, and on the right is the concept implemented by PP250. Each idea builds on the prior concept, increasing functionality while minimizing complexity and staying true-to-form and true-to-function. The λ-calculus binds a limited but fully functional digital working space for each calculation frame as a binary computer (Turing's α-machine).

A FUNCTIONAL CLASS

The Abacus only modelled arithmetic, and it took Millenia to reach step two. That started in 1614 with Napier and matured in 1622 when the algorithmic scale of his logarithmic expression was embedded as the mathematical function of the machine. A slide rule is pre-programmed, and the algorithm no longer needs to be learned as with the Abacus or Napier's Bones. The machine can perform more than one function. The algorithm is just another λ-calculus Variable. The generic machine adds two symbolic lengths from the logarithmic scales on a shaft and a sliding rod. This dynamic action of the universal model of computation substitutes the numeric variables and selects the programmed scales to frame the thread of computation.

The Slide Rule is a *'Class'* of object-oriented machinery with different functions implemented on independent scales. This time, the DNA string of the design as a blueprint for the machine lists each function abstraction bound to the computer, such as multiply, divide, square, root, tan, sin, etc. The Class is a generic object-oriented framework for the computation of more than one related function. With a Slide Rule, the numeric Variables are bound to the chosen function by a cursor using the human eye. However, the scales, the sliding mechanics, and the human eye mean errors can still occur.

As Babbage knew well, human error is a curse to computer science, even if a logbook is used. The source of the logbook is subject to human errors ranging from wear and tear to mistakes and even sabotage, while the Slide Rule is dynamic for mathematical calculations made on the fly.

The ultimate step removes all sources of human error. It was demonstrated by Babbage circa 1840 at the pinnacle of the Industrial Revolution. He removed all sources of human error, leaving only mathematics. He constrained the variables to a chain of mathematical function abstractions linked end-to-end as symbols written on a chalkboard at school, in a university, or on a pack of twenty-five punched cards. Visibly, mathematics weaves this DNA on the chalkboard of expressions. It is how Ada Lovelace programmed her perfect solution to Bernoulli Numbers as a function abstraction. Babbage went as far as to automate the printing of tabular results to avoid the printer's errors when typesetting over 10,000 numbers to more than fourteen decimal places.

Babbage's proof of *Infallible Automation* remains the Holy Grail of computer science. The universal model of computation, the λ-calculus and Capability Limited Addressing are the three essential ingredients to free digital computers from human threats and errors. It requires Alonzo Church's dynamic λ- calculus meta-machine to keep Turing's procedural software on track.

When software is structured mathematically as object- oriented machine code for the universal model of computation by a Church-Turing Machine, a proven (evaluated) expression cannot be silently attacked and never needs to be urgently patched. Service is always fail-safe and restored at once through *Infallible Automation*. All software efforts can focus on functions, improvements, and evolution. Industry, business, government, and citizens control their confidential data using their secret Capability Keys in their private Namespace.

THE SURFACE OF CYBERSPACE

T he surface of computer science is the wild digital frontier where all action takes place. It is the dead centre of the Church- Turing Thesis, where software interacts with hardware on the distributed surfaces of cyberspace. It is where the λ-calculus encapsulates the imperative Turing-Commands to unfold the flawless λ-calculus frames as atomic calculation.

The appointed context registers scope the imperative commands of Turing's α-machine to crosscheck and balance programs against the DNA strings. It filters out bugs and malware, and it locks out hacking. The Capability Keys control the cybersecurity policy, not only on this digital frontier but also through the directional DNA hierarchy into the hands of the Namespace owner.

At every step, Turing's α-machine is sheltered as a mathematical framework for a calculation that obeys the same variable substitution rules learned at school. The substituted λ- calculus Variables define the values, the class of abstraction, the class function, the thread, and the namespace table. It should be noted that Capability Keys equally constrain network communications. Thus, a distributed denial-of-service attack is automatically avoided since the TCP/IP address is hidden by the point-to-point Capability Keys. Everything is pre-approved or denied by the mathematics on the chalkboard translated as a DNA blueprint.

The cockpit of the computational gene is a binary computer (Turing's α-machine), harnessed to Church's λ-calculus meta-machine. It is where software ideas turn into actions. Given the technology stack of a Church-Turing Machine, boundaries cannot be bypassed, and hackers have no unfair or unregulated privileges to use. Nothing is hidden, opaque or remains undetected. Every step is authorised and crosschecked. A dormant, unrecognised program bug is consistently detected on the first occurrence without causing undetected harm or hidden corruption.

TRANSPARENT CYBERSECURITY

While mathematics and every function in society can be virtualised, security must be transparent because it is a property of physics. Controlled by Capability Keys in the hands of the Namespace user, cybersecurity is under user policy control. The Church-Turing Machine has no power or authority unless granted and delegated by the user who controls the Namespace as a software machine. The two-dimensional substitution mechanisms in the universal computation model separate data from functions. The citizens control data privacy through transparent Capability Keys, which distributes power as a cyber democracy in the independent hands of society. The qualified service function can be imported from a competitive marketplace without the threats of General-Purpose Computer Science.

The universal model of computation reverses and ends the mistakes inherited from the pre-electronic age. General-purpose computer Science's baggage is removed and replaced by the science of the λ-calculus and Capability-Limited Addressing: the mathematics and the transparent means of cybersecurity level the playing field for anyone to become a programmer and a computer expert. No opaque privilege like Main()[84] exists. Its unworthy approach does nothing to defend software from cyber- attacks. Instead, calculations are controlled by the mathematics and capability keys needed to compute, and all are administered privately.

---

84   Main() is a special function for the starting point of a program. The main function is not predefined by the compiler, it is supplied in the program text to install

Virtualised society must coexist in the electronic village of the 21st century with the third world, with communist dictatorships and competitive states directly linked to the beating heart of industrial society. For democracy to survive and prosper, digital boundaries are vital, and cybersecurity must be controlled by individuals and communities, not monopolistic, dictatorial suppliers.

The international fuss over Huawei and TikTok technology exemplifies the problem. If computers used the Church-Turing Architecture, every point-to-point communication link could be privately encrypted to avoid the issues of spies and snoops of any political persuasion. To liberate cyber-citizens of the 21st century, function-by-function, point-to-point encrypted connections will be an essential part of a free and functioning cyber democracy.

---

the operating system kernel and setup the superuser.

THE λ-CALCULUS CONCEPTS

Type-sensitive Capability Keys and type-sensitive Church Instructions match to unlock the λ-calculus Variables, Abstractions, Functions, and Applications in a Namespace as computational types. The type of sensitive Church Instructions engages the specific cogs and leavers of a clockwork λ-calculus engine that controls the context of frames as a software machine of mathematical symbols. The symbols define media-tight separation within the universal model of computation as a software Namespace in a Church-Turing Machine.

The Namespace table relates the two sides of the Church- Turing Thesis. Each new Capability Key is minted with embedded access rights that add a slot to the Namespace Table and then return the Capability Key to the owner in a context register as a new digital object that belongs to the Namespace. Each Capability Key is a token or symbol with a name and a digitally assembled typed and access-limited object. The status and details of the in-form geometric location or the out-form networked location are stored in the Namespace table.

Each function abstraction is defined by a C-List storing the DNA string of Capability Keys to sub-atomic and external function abstractions. The blueprint of one node in the directional DNA hierarchy limits the scope of the computation. It also restricts the movement through the directional DNA hierarchy. An accessible C-List has 'Load' or 'Save' access rights to the DNA list of Capability Keys. The strings of DNA are linked by 'EnterOnly' access rights that can only be 'called' to unfold a new frame of computation.

The *'EnterOnly'* access right allows one node to link to another in the directional DNA hierarchy, but this unique access right prevents interference since *Load* or *Save* are missing. The call Church Instruction microcode adds *Load* access right to the context register CR[6] if the transfer of control is successful. Please note that this temporary permission has not been added to the Capability Key. Thus, as one computation frame *'suspends'*, a new frame is *'entered.'* The Capability Keys and the Church Instruction dictate the switchover protocol to unlock and unfold the chosen link in a chain of independent, fail-safe, functional computations.

Syntax	Church Instruction	Λ-calculus Description	Clockwork Meta-Machine
a, b...+,- ...	Variable, **Load,** or **Save** Capability Key to Thread Context	A Capability Key as a Variable	The Variable 'a'... is an Immutable Capability Key showing the digital geometry of the λ-calculus Variablewith a type-limited access right unlocked by a '**Load** or **Save** Capability' micropro-gram.
(λx.M)	Function Abstrac-tion, **Call** and **Return**Selecting Function Context	Function as an exe-cutable Capability Key bound to a Class asan *EnterOnly* Capability Key	Synchronous **Call-Return** protocol.M as a Λ-term is an *EnterOnly* Capability Key to an Abstraction implemented as a C-List to scope the subatomic media of the Object-Oriented Class and pick the executable function as binary code from the C-List.
(M N)	Application, **Change** Thread Context (Suspendand acti-vate newThread)	A Thread M as an *EnterOnly* Capability Key carrying event variables N as Capa-bility Keys to Function Abstractions.	**Change** the Parallel Thread protocol. N is an *EnterOnly* Capability Key to a Thread, a complex C-List with all Computational State Words. It includes Capability Keys to in-use context registers to list the symbols' access rights for ThreadM.
E	Namespace, **Switch** Name-space andChange Thread	Replication case of Λ-Calculus Name-space Theoryas Capability Key E	**Switch** the network protocol. *Enter-Only* Capability Key to a Namespace Table and run the default Thread. The Namespace lists the geometry of the objects inthe Namespace E.

FIGURE 15: TYPED Λ-CALCULUS CONCEPTS VS. CAPABILITY-LIMITED CHURCH INSTRUCTIONS

These details are covered later, but the six Church Instructions are highlighted in bold in Figure 15 (Load, Save, Call, Return, Change, and Switch). They are the gear sticks that unfold the computation framework for Turing's α-machine. The woven mathematical structure is revealed by these hidden programmed springs.

All errors detected on the spot by Capability Limited Addressing are treated as an impending attack. Immediately, the Thread is aborted, and the Church-Turing Machine 'Switches' to invoke the Recovery Namespace and start a fresh instance of the Recovery Thread. Its instinctive, hardware-driven reaction is the fail-safe, nervous response to any error. The aborted Thread identifies the invalid step with the referenced Capability Keys to diagnose the root cause and report back. Diagnostic tests isolate hardware errors and update the MTBF of implicated software objects for orderly improvement.

THE TYPED ACCESS RIGHTS

D irect access to a sub-atomic binary object is strongly typed, while access to objects in a C-List is always indirect. Indirect access hides the implementation as an enclosed digital black box with locked and gated access rights. Its framework safely shapes each computation from the decimal rail of an Abacus to a logarithmic scale of a Slide Rule and the mathematical expressions of Ada's function abstraction to a telephone switching function or a legal document in a virtualised society. Transparent security rules are enforced within and between objects by every computation frame. Access to any digital item requires the object to be first unlocked when a Capability Key is inserted in a Church Instruction. Its conscious action transparently unites security with functionality. Thus, Capability Keys from the directional DNA hierarchy add a second mechanism of programmed control as the λ-calculus Variables are substituted into the computation framework.

The core of the microcode is the Load Capability and Save Capability micro-sequence that unlocks access rights, installs boundary checks, or updates a C-List with a locally substituted Capability Key. Any hardware trap conditions are resolved according to the logical and physical conditions concerning the Capability Key status in conjunction with the Namespace slot status, including any network location.

The 25-bit word included an address-and-data parity bit and limited the access rights to 8-bits for PP250 when the 16-bit word was typical. It covered critical needs, but a redesign with a 32-bit word would be a helpful improvement. More than this is optional since decomposition for digital security makes virtual memory both unwanted and unnecessary. Scaling atomic abstractions as an atomic network is far more valuable for less effort and cost.

To emphasise this point, abstracting virtual memory as a page-based function abstraction is far easier to customise since no special hardware is used. Its ability to abstract any General- Purpose Computer in a Church-Turing Machine validates the infinite power of a Church-Turing Machine over a General- Purpose Computer. It also smooths the migration to phase out the pre-electronic age and speed in the electronic age architecture for the future. Different word lengths, from small to large, create competition. PP250 uses a microcode escape mechanism to extend the microcode with function abstractions seamlessly. For example, a trap condition in a context register hides a frame change to a function abstraction that resolves the trap condition. The function abstraction could activate an asynchronous service as a local or remote call to a networked namespace.

Every action is limited by the need to know and limited access rights field of each Capability Key. Figure 16 details the format of a PP250 Capability Key. The access rights are cached with the location scope in each context register. Its media-tight security system frames every bound computation object in a hardware shell.

Typed Access Rights (8 bits)							16-bit Key Identity - Slot in Namespace (The 72 bit Slot Entry holds the local or remote cyber geometry of the key)	Address + Data Parity Check Bit
T	E	S	L	X	W	R		
↓	↓	↓	↓	↓	↓	Read Data	Binary Turing Types	
					Write Data		Read and/or Write Data or	
				Execute Program			Executable Code	
			Load Capability Allowed				C-List Church Types Load, Save	
		Save Capability Allowed					Capability Keys or *EnterOnly*	
	EnterOnly Allowed						Black-Box frames.	
00 In-Form Key (Normal {*AccessRight.ObjectID*}) 10 Out-Form Key ({*AccessRight.DiscLocation*}) 01 Passive Key (22-Bit Interworking Token) 11 Literal Key (72-bit Sheltered Binary Value)							Trapped Binding Attributes resolved by Interrupt on use or load	

FIGURE 16: ACCESS RIGHTS IN IMMUTABLE CAPABILITY KEYS (PP250)

A Last-In-First-Out (LIFO) stack frames a Thread and caches each frame by pushing and popping the C-List, the Capability Keys to the program and the next instruction offset as a breadcrumb trail to return later. These two Capability Keys are reloaded and re-synchronised using a Return() instruction.

THE FRAMES OF COMPUTATION

An *EnterOnly* Capability Key identifies a functional *'frame'* as a function abstraction, but the DNA string in a C-List is hidden until activated by the calling frame. The C-List defines the root node of a string of DNA found in the namespace. Each frame is represented by an *'EnterOnly'* Capability Key to a C-List. The new frame has one of three forms. They all start the same way as a function abstraction using an *'EnterOnly'* Capability Key.

- First, a sub-routine call occurs within the current Thread context. The *EnterOnly* Capability Key defines the Class of object-oriented machine code as an abstraction, and the instruction's offset selects the *'Execute'* guard program of the Class. The other context registers pass the λ-calculus Variables and any other data parameters.

- If the next frame is asynchronous, a parallel Thread can be activated by unfolding the Capability Keys from the DNA string. The context must include the Thread to activate. The Dijkstra flag abstraction organises it.

- If the next frame is remote, the remote Thread is activated by a network request to unfold the Capability Keys in another Namespace. A Passive Capability Key identifies the remote namespace. Once again, this can be coordinated by the Dijkstra flag abstraction.

The Passive Capability Key is an immutable value translated by a function abstraction for Passive Capability Keys, which finds and connects to the remote namespace to run the function. Once again, a Dijkstra callback flag abstracts a shared, point-to-point communications channel to synchronise parallel Threads. Local and remote threads can be grouped and synchronised by an ordinary transaction. Point-to-point communication channels can be individually encrypted to prevent man-in-the-middle attacks.

In each case, the hidden springs unfold their DNA, identified by an *EnterOnly* Capability Key that hides a framework of object- oriented machine code. Only the microcode can access an *EnterOnly* Capability Key to change frames—the Call-Instruction aborts any error detected during the validation and verification steps.

A computation includes periods of waiting for an asynchronous event to mature. When reactivated by the asynchronous event, the context is rebound from the immutable Capability Keys. It synchronises the context registers with any memory changes and coordinates with any active garbage collection requests concerning outdated Capability Keys as indicated by the status of the slot in the Namespace Table. The Namespace allows objects to enter or leave the memory space or move in cyberspace without interfering constraints. The dynamic binding of the λ-calculus meta-machine separates each independent problem space within the network. The universal computation model and the λ-calculus as a meta-machine dynamically bind every resource, and all caveats are scientifically resolved.

Memory location changes in a network are required, and the updated Namespace table supplies a simple solution to a common problem. Whenever a Thread restarts, the reloaded Capability Keys automatically synchronise the context registers as an in-form or out-form memory segment.

Further, any variable can be a binary value of one or more of the eight data accumulators or a complex function abstraction as a Capability Key in a context register. A variable returned by a function call is an anonymous λ-calculus variable until it is 'saved' in a named location of a C-List. It can be delegated without difficulty or unexpected side effects that plague the static memory of a General-Purpose Computer and with all the additional power of the λ-calculus and anonymous functions run by high-performance machine code without baggage or threats.

For example, when telecommunication software constructs a voice connection between any two phones worldwide, a DNA string defines this connection as a global mathematically guaranteed structure named locally *'MyConnection.'* The Capability Keys to the complex structure and peculiar details of any telephone connection using multi-vendor switching systems are standardised as a DNA string of locally substituted, functional Capability Keys representing an infinite set of variables. It dynamically configured directional DNA hierarchy implemented by local and networked Capability Keys strictly, safely and securely reaching right around the world.

The Internet today is the same global, multi-vendor, ever- evolving problem. It can only be solved comprehensibly and in full using the universal model of computation and locally substituted capability keys where functionality and security coexist side by side, atomically. The monolithic, static compilations of General-Purpose Computer Science have no chance to solve all these problems simultaneously across an ever-changing global network. Safe access to the *'Internet of Things'* must be limited by a need-to-know and the least-privilege rules of security. Protecting the Internet of Things as function abstractions in Church-Turing Machines spreads networked cybersecurity to each physical object, protected from malware and hackers, from denial-of-service attacks and spying. These, the unsolvable problems of General-Purpose Computer Science, must be resolved for civilisations to survive and progress.

THE λ-CALCULUS VARIABLES

The essence of λ-calculus is rooted in the universal model of computation; it underpins a scientific process for stable evolution. The problem is solved atomically in nature and by a Church-Turing Machine for computer science. The dynamic binding of a λ-calculus meta-machine separates each scientific concern as an independent issue. It allows the problems to be solved independently and safely in depth and detail.

Multiple problems are solved independently as hidden springs of a nuclear framework. A nuclear framework is functionally stable yet adapts to physical and functional changes. At the same time, the details are hidden as atomic components. Functional symbols define the interrelated but independent solution to natural problems. A physical map is a skeleton of locally restricted functions that implement the nuclear blueprint of a dynamic, survivable, evolvable, working species.

This machine originated as the Abacus designed to abstract just one functional class. It is an object-oriented machine that has parts. Using atomic replication scales the parts as a nuclear machine for large decimal numbers.

In every case, the DNA of atomic type limited relationships are passed from generation to generation as the theory of some complex functions bound together by the laws of λ-calculus. The Babylonians discovered the mathematical framework required for decimal arithmetic.

Bees used natural selection to find another structure at a far earlier time. Their six-sided honeycomb is the most effective framework for survival and scalable growth as a beehive. The social organization of bees in a hive can teach society how to share cyberspace as a functioning democracy which can survive. More is explained in Chapter 13 – On Digital Enlightenment.

Natural selection uses the universal computation model to drive everything in nature from pure mathematics. The Church-Turing Thesis captures the universal model of computation utilising the λ-calculus as the scientific foundation that holds the different pieces together as the machinery of ever-evolving yet stable species.

It starts with the components of expressions named λ- calculus Variables. Consider the object-oriented abstractions of an Abacus shown in Figure 17. The expression is valid for a single digit between zero and nine, but the Abacus scales to any large decimal number as an array of individual abstractions through replication. The array performs addition using the same atomic functions.

$$a + b = c$$

or for subtraction

$$a - b = d$$

FIGURE 17: THE MATHEMATICAL FUNCTIONS OF THE ABACUS

These two functions, *add(a,b)* and *subtract(a,b)*, are implemented by two methods in an object-oriented mathematical class. In a class, the function, as a class method, is also a λ-calculus variable, substituted into the 'single-tape reader' (CR[7] for PP250) on a case-by-case basis as required.

Likewise, a method is dynamically chosen in a Slide Rule. It is the same in Turing's α-machine when the tape is changed or in a Church-Turing Machine when a Capability Key reloads the context registers as an object-oriented symbol. The laws of the λ-calculus infallibly automate its selection mechanism. A λ-calculus meta machine changes the framework of a class by the infallible automated science of the λ-calculus. A directional DNA hierarchy of immutable Capability Keys defines chains of object-oriented machine code. Capability Keys substitutes executable algorithms, the exact mechanism in every respect, as any other λ-calculus variable substitution.

A computation frame is reduced and simplified to Turing's α- machine dynamically bound to λ-calculus variables—nothing more and nothing less than the context registers in Figure 12. A network of Church-Turing machines switching between their frames of computation, executing multi-dimensional nuclear strings of engineered functionality.

Thus, algorithmic functions and other variables are substituted by the exact generic Load-Capability mechanisms applied to Variables, Abstractions, Functions, and Applications. They are all just symbols like the numeric variables *(a, b, +* and *-)* used at school. The computational formula is always the same simple Turing α-machine, but the context registers in Figure 12 change as the DNA is traversed. Turing's α-machine is digitally reshaped by different run-time values substituted by the directional DNA hierarchy into the set of context registers.

Indeed, in the above examples, each scientific symbol from Figure 17 *a, b, c, d, +, -* and = all have specialised substituted implementations in the calculation. It is a generic clockwork process of substitution within and between the chained frame of functional computation. It is the symbolic half of the universal model of computation. The other half is a digital implementation of a mathematical calculation performed in splendid isolation by the latest technology, a binary computer acting as Turing's original α-machine.

In turn, the frames are bound to Turing's α-machine by the λ-calculus meta-machine, following the DNA blueprint mechanised by Church Instructions and the scope of Capability Keys in nodal C-Lists. Unlike labelled storage in statically bound virtual memory, the object-oriented machine obeys the Babylonian Golden Rule. The objects gain value as a trusted asset calibrated over time. The Capability Keys is the handle of an unchanging object with a proven value that can only increase over decades of stable service. The storage location is immaterial.

For example, as a network telecommunication switching system, the expression in Figure 18 can be evaluated by threads in the same and different machines without any of the baggage needed by statically compiled General-Purpose Computers. Four Capability Keys define the local implementation dynamically bound as a unit of context by a single, indivisible, machine code Church Instruction.

$$myPhoneCall = Phone.\ ConnectMeTo(myDoctor)$$

FIGURE 18: A FUNCTIONAL CHURCH INSTRUCTION IN A TELECOMMUNICATIONS NAMESPACE

This high-level machine command is characterised by the framed context of well-chosen names to the Capability Keys. Each Thread sensitive context in a private Namespace solves the λ-calculus Variables *myConnection* and *myDoctor* with local values while the function abstractions *Phone.ConnectMeTo()* is a generic object-oriented machine code a service provider offers. The computational context of each Thread dynamically drives binding.

The Capability Keys increase performance by reducing the executed code. In a loop, a dynamically delegated Capability Key acts as an anonymous function at the machine level.

COMPUTATIONAL THREADS

The symbol '=' has a special meaning of 'calculate', at which point a functional Church Instruction evaluates the substituted variables and later returns a result. The evaluation transfers Thread ownership to an *EnterOnly* frame using the chosen function. The initial step is a synchronous, Last-In-First-Out (LIFO) subroutine Call. The Thread stack saves the return context and unfolds the new framework that becomes the context at the top of the Thread stack. Its synchronous evaluation suspends the calling function abstraction while another frame calculates and returns the result.

Using a Church-Turing Machine for computation, a single Thread calculates and returns results using a locally bound object-oriented function abstraction passing λ-calculus Variables within the active Thread. No centralised operating system or statically privileged virtual machines are needed. Even the functions of a store manager or any device driver can be activated in line private, fail-safe abstractions. The frame-based security of Capability Keys does all the hard work of isolating critical function abstractions from each other. As a result, the centralised privileges of General-Purpose Computer Science, the overhead, and the delay are all replaced by object-oriented machine code and functional Church Instructions.

In addition to the Church Instruction for a synchronous call, two other framework-changing Church Instructions exist. First, suspend a thread and change it to another, then swap it to a remote or local namespace. The exact microprogrammed mechanism is also used for error recovery. The more complex cases are programmed as system abstraction managers that replace the superuser in a General-Purpose Computer. For example, a parallel Thread is primed by a Thread manager before it runs, and a Namespace agent organises a call to another or a foreign Namespace to activate a remote functional Thread.

CHAPTER 9

Infallible Automation

The chief drawback hitherto on most of such machines is, that they require the continual intervention of a human agent to regulate their movements, and thence arises a source of errors; so that, if their use has not become general for large numerical calculations, it is because they have not in fact resolved the double problem which the question presents, that of correctness in the results, united with economy of time.
Luigi Federico Menabrea 1809 - 1896, author of 'Sketch of the Analytical Engine,' first Prime Minister of Italy (1867 - 1869), a General, a Count, a Marquess, a statesman, a mathematician, and a friend of Charles Babbage

In 1935 and 1936, when Alan Turing joined Alonzo Church at Princeton University, the theory and practice of digital computer science crystalised Ada's mathematical dream from a century earlier. In a direct lineage, Babbage's *Infallible Automation* descends from the Abacus, Slide Rule, his proposal for the Analytical engine, and Ada's vision as an Anthem for the future of Computer Science. Alonzo Church perfected *Infallible Automation* as the universal computation model and the λ- calculus laws.

At the same time, he guided his student Alan to his single-tape Turing α-machine that exemplifies Alonzo's λ-calculus Application. Turing's α-machine is a digital gene to compute Alonzo's mathematical DNA in universal cyberspace. The two halves of the Church-Turing Thesis came together as the science of *Infallible Automation*. Infallible automation began a century earlier in the heyday of the Industrial Age with Charles Babbage's two mechanical engines. In 1833, Ada Lovelace, the curious, creative daughter of the poet Lord Byron, met Charles Babbage. Ada, a skilled mathematician in her own right, was inspired to dream.

After Babbage lectured in Turin in 1840, Charles Wheatstone asked Ada to translate Luigi Menabrea's notes. With Babbage's encouragement, Ada added something to her dream. Like Luigi before her, Ada recognised Babbage's second engine, the Analytical Engine, as revealing the future, a flawless, mathematical future without the scourge of human errors and failings. She translated Luigi's memoir and added her vision for solving nature's most profound mysteries through sound computer science.

> *'Supposing, for instance, that the fundamental relations of pitched sounds in the science of harmony and of musical composition were susceptible of such expression and adaptations, the [Analytical] engine might compose elaborate and scientific pieces of music of any degree of complexity or extent. The Analytical Engine is an embodying of the science of operations, constructed with peculiar reference to abstract number as the subject of those operations.'* A. A. L.

Babbage's first engine was hardwired as clockwork functions to flawlessly compute and print polynomial tables from pure mathematical expressions without human error, for example, the quadratic polynomial.

$$p(x) = 2x^2 - 3x + 2$$

The Difference Engine calculated, tabulated and printed the results as a flawless table for the values *p(0), p(1), p(2), p(3), p(4),* and so on. Babbage was determined to avoid all human errors, including printer errors. His mantra for computer science was monumental: *Infallible Automation.*

Mathematical perfection should be the talisman for all engaged in computer science, but sadly, it is not so. Digital dictators refuse to perfect computer science for the benefit of others. They will always remain self-interested monopolies. Babbage's computers could calculate and print perfect results to many decimal places as requested without human involvement. Disappointingly, the human errors that prompted his mission have returned. Malware and hacking pervade General-Purpose Computer Science in force from every enemy of democratic civilization.

Ada described Babbage's flawless working prototype in 1833:

> *'We both went to see the thinking machine*
> *(for so it seems) last Monday. It raised several*
> *Nos. [numbers] to the 2nd and 3rd powers and*
> *extracted the root of a Quadratic equation.'*
> *A. A. L.*

However, soon after this impressive demonstration, Babbage suspended work on his Difference Engine. His ideas had matured into the Analytical Engine that did much more than his hardwired and statically limited Difference Engine. The analytical engine was mathematically programmable, no longer hardwired to just a polynomial formula, such as incomprehensible nuts and bolts or misleading claims that the nuts and bolts are general-purpose.

His computer was applied science engineered as a mathematical machine. The machine language was simple mathematics woven into a complex calculation. Any values or symbolic expressions could be stored and later combined algebraically or arithmetically in endless ways to resolve all mathematical complexity. For this, Babbage decomposed the structure he used for polynomials into the *'four operations of simple arithmetic upon any numbers whatever.'*

THE ANALYTICAL ENGINE

The Analytical Engine is a perfect example of industrial-strength, trusted computer science. Luigi Menabrea (1809-1896), a mathematician before he became a General, *'grokked'* this at a meeting with Babbage in Turin. He was so impressed that he wrote and published a 'sketch' of how the Analytical Engine worked in French. Translating from this earlier work, Ada Lovelace updated Luigi's explanatory documentation with an extensive seven-part appendix[85]. In *'Notes A to G,'* Ada explained her vision beyond just programming the Analytical Engine. Her notes build up to Note G, where Ada explains in-depth how to solve the numbers in a Bernoulli series as pure mathematics, what is today called functional programming.

More importantly, but unrecognized and undiscussed, in Ada's program for the Analytical Engine, she wrote mathematically and passed variables as functions using the rules of a λ-calculus function abstraction.

'Expressing the ratio of the circumference to the diameter, the Numbers of Bernoulli, &c., which frequently present

---

85   Sketch of The Analytical Engine Invented by Charles Babbage By L. F. Menabrea of Turin, Officer of the Military Engineers from the Bibliothèque Universelle de Genève, October, 1842, No. 82 With notes upon the Memoir by the Translator Ada Augusta, Countess of Lovelace (found here https://www.fourmilab.ch/babbage/sketch.html)

themselves in calculations.'

Her algorithm was published in 1843 within Note G, elucidated on a single page with full explanations. Her twenty- five mathematical steps as 25 statements compute the Numbers of Bernoulli using Babbage's Analytical Engine. While this program could not be run on a General-Purpose Computer, it could, after two hundred years, still be executed as a flawless function abstraction on a Church-Turing Machine. Flawless mathematical software that lasts forever will change the destiny of humanity and civilisation. Beyond cybersecurity alone, it is a profound and overwhelming reason to replace outdated, centralized binary computers with Church-Turing Machines.

Industrial strength in computer science will elevate civilization to another level of productivity, science, and civilized democracy, where industrial strength software will last forever and be transparently readable by anyone interested. Cyberspace will be a digital foundation for all without branded experts, undetected cyber crime, and digital dictators. The flawless function abstractions of cyberspace will enhance society in ways beyond Ada's vision that we cannot yet see.

A General-Purpose Computer does not understand functional mathematics written on a school chalkboard. For example, Ada's program includes a loop of ten precise steps (see Figure 23). As Ada wrote, she commented on her vision of the infinite and endless future with perfect, fail-safe computers to decode nature's secrets and every other interest of humanity. At the start of this book, I quote her words as an anthem to computer science.

Ada's *'algorithm'* as she explained in Note G, is considered the first computer program; together with the text, her vision of endless, flawless science, using mathematics as a universal, perfect machine language *'to express the great facts of the natural world,'* including as she wrote music, philosophy, and nature. Her vision of a flawless machine to solve all the mysteries of science and nature puts her on the same level as Steve Jobs and his vision of usability through object-oriented programming.

When object-oriented programming becomes object- oriented machine code, the Dream Machine results. It is expressed through perfect functional Namspaces for mathematics and anything else, infallibly automated and free from undetected human error and interfering mischief. Ada and Babbage used mathematics in letters, which they exchanged almost daily in the summer of 1843[86].

The same expressions are written on chalkboards worldwide and then resolved by students with flawless results. One might reasonably ask why von Neumann used incomprehensible binary machine code, forcing compilers and programming languages when he should have adopted Babbage's better idea, where the Namespace symbols become the programming language.

As a pure mathematical machine, no awkward compiler is needed. Babbage wrote all the machine code himself as the cogs and leavers for the *'four basic functions'* built into his machine. His meta-machine was a combination of the Abacus and the Slide Rule, programmed by pure mathematics, as Ada explained using the more *'interesting'* expression in Figure 19.

$$\int \frac{x^n\, dx}{\sqrt{a^2 - x^2}}$$

FIGURE 19: INTERESTING EXAMPLE FOR FOLLOWING THE PROCESSES OF THE ANALYTICAL ENGINE DOCUMENTED BY A.A.L.

---

86   Ada, The Enchantress of Numbers, A selection from the Letters of Lord Byrons's daughter and her description of the first computer, Narrated and Edited by Betty Alexandra Toole.

The use of flawless mathematics avoids all human-inspired errors. It is a scientifically level playing field, a machine as faithful for lay citizens as for experts. There are no unfair privileges for hackers or criminal malware, no superuser administrator or centralized operating system, and no way to disrupt mathematics. In short, *Infallible Automation* is Babbage's hallmark for preventing human error. It led Ada to her vision of a universal computer programmed beyond pure mathematics.

There can be no malware or hacking if computers obey the λ-calculus. Its machine language is impervious to bugs, malware, and hacking. Internal bugs are found on the spot. Flaws cannot upset the process, while λ-calculus holds the chalk and the eraser. The computations obey. As implemented by Babbage and the λ-calculus, applying science and technology removes human errors through *Infallible Automation*.

Thus, a mathematical expression on a chalkboard that students reduce by hand has clear rules that mechanise a flawless, fail-safe process, a clockwork machine. Ada wrote her program this way, as pure mathematical expressions. As she wrote in her example, see Figure 19 taken from Note G, she dreamed of a flawlessly automated machine exemplified by Babbage's working prototype of the Difference Engine, engineered mechanically, infallibly automated symbol by symbol from start to end, to avoid all human error[87].

---

87   As the story goes, in 1821 Babbage and his friend, the astronomer, John Herschel, were checking manually calculated tables. Babbage found error after error, and soon exclaimed '*I wish to God these calculations had been executed by steam*'. The grindingly tedious labour of manually creation and double checking was one thing, their unreliability was another. Thus, he embarked on an ambitious engineering task to design and build mechanical calculating engines of unprecedented size and intricacy, to eliminate all human error. His infallible machines eliminate both the risk of calculation errors and transcription when copying the results by avoiding manually setting results as loose type. Stereotyping automatically impressed results on a soft material as a printing plate and prevented printing errors. The outcome would be flawless as was his intention.

The machine code for *'four operations of simple arithmetic upon any numbers whatever'* was designed in-depth and detailed by Babbage, and no reprogramming was needed. A Church- Turing Machine is the same.

No need exists for human practices beyond the ability to write mathematical expressions. No branded, dictatorial operating system and proprietary, binary conventions are needed to confuse. For Babbage, perfect results without any risk or threat of human error. In 1936, the universal model of computation scientifically combined Ada's dream with Babbage's ideal machine, and digital computers began.

However, a binary computer proposed by von Neumann and sold today by digital dictators cannot support symbolic programming in machine code. For this, object-oriented machine code and Capability addressing are required.

ALONZO CHURCH

Alonzo considered the universal model of computation as nature's engine room. Here, the λ-calculus frames the algebraic chains of dynamically bound, object-oriented calculations. The arrival of Alan Turing at Princeton University turned his theory into practice.

Using the limited technology of the day, Turing outlined his single-tape α-machine to apply variables to a programmed function encapsulated on one paper tape. It is what Alonzo Church called an '*Application*.' Turing's α-machine encapsulated a computation exactly as Alonzo Church proposed for the λ- calculus. Turing's α-machine computes one algorithm with one set of variables precisely as the λ-calculus requires.

Turing's single-tape α-machine perfectly matches Alonzo's theory as a universal model of computation that scales atomically as nature's solution to dynamic calculations by living organisms. The universal model of computation is the prototypical form of a Church-Turing Machine.

Alonzo Church is the unsung hero of computer science. He was born in 1903, the son of a judge in Washington D.C., and by twenty-four, he had earned a doctorate from Princeton University. He worked as a research fellow at Harvard, Göttingen in Germany, and Amsterdam. When he returned in 1929, he was appointed Associate Professor in Mathematics. Princeton University was the hotbed of *'logic,'* and besides Church and Turing, Rosser, Kleen, von Neumann, and Gödel worked or spent time there. Since then, Alonzo's research on mathematical logic, replication, and recursion theory has only increased in value.

Programming languages have made dramatic improvements using theories from the λ-calculus. Implementing the λ-calculus as machine code is the next step for civilised scientific progress as Industrial Strength Computer Science.

His search for λ-calculus, his most fascinating discovery, began in 1930. By 1936, the same year Alan Turing published his research *'On Computable Numbers, with an Application to the Entscheidungsproblem,'* Alonzo published his results, his holy grail. The λ-calculus formalise his vision of an endlessly linked mathematical algebra using a universal model of computation. His scientific symbols are defined in chains of mathematical DNA. When implemented as capability-limited object-oriented machine code, the theory and practice match the dead centre of the Church-Turing Thesis on the surface of computer science as actions form and take place.

Ada's dream from 1845 was codified as pure chains of symbolic mathematics. The form of symbols exactly matches an expressed function. United by three basic rules (see Figure 14) and a minimal number of concepts, the Variable, Abstraction, Function, and Application in a λ-calculus Namespace. Function abstractions in an infinite variety of specialised forms hide their functions as Bronowski's coiled but hidden springs in nature. A directional DNA hierarchy defines a dynamic, stable, evolving functional species.

The universal model of computation starts with the λ-calculus as a Namespace representing the DNA of a living species, either natural or digital. In the digital world, cyberspace is populated by various species, each a private Namespace. Object-oriented machine code fills the namespace structured by a DNA of Capability Keys.

Think of a Namespace as a digital mirror to the chalk symbols on a university chalkboard. The symbols are, at once, the λ- calculus Variables and the digital atoms of computer science. They are not simply values.

A λ-calculus Variable is a name for the unchanging handle to a mathematical or functional concept, a specific function abstraction. As immutable tokens of civilised power, like paper money, Capability Keys tame the digital markets in cyberspace. By representing sophisticated, minted tokens of guaranteed significance, Capability Keys easily carry digital structures of extreme complexity as simple event variables to the functional organs of a local or networked Namespace.

PP250 used this power to simplify the global telecommunications network. Programmers programmed in high-level object-oriented machine code, Namespace by Namespace, function abstraction by function abstraction, object by object, and fail-safe instruction by fail-safe instruction, without using any distraction. No indifferent programming language, monolithic compilation, centralized operating dictator, outside interference, or complex security strategy was considered. When flawless digital security is built in, everything to do with software is simplified.

Alonzo's symbols obey the Babylonian Golden Rule of form-matching function. It is a feature of abstraction to reduce and remove unnecessary baggage, leaving only the essence behind. When the λ-calculus variable's name captures that essence, the high-level programming language is designed automatically to empower statements like:

$$myConnection = SwitchNewZeland.\ Connect(me, myMother)$$

FIGURE 20 A PRIVATE TELECOMMUNICATIONS NAMESPACE EXAMPLE

The λ-calculus variables in any Namespace compose the language's syntax and semantics by naming the Capability Keys that direct the thread computations as object-oriented machine code.

Computations in the universal model of computation are nuclear threads. A Thread as a λ-calculus Application carries an event in the form of λ-calculus variables, such as an event abstraction of a telephone number from a smartphone to other function abstractions that process this specific event variable, together with conditions like *off-hook, busy, out-of-service* and so on. There is no centralised operating system in PP250, in the λ-calculus, or nature: species, organs, and abstractions as components function atomically, moving and communicating point to point.

Each symbol is an expression of functional logic. The symbols on the chalkboard hide the details. Consider the most straightforward calculation first learned at school: *a = b + c.* It is also the expression first converted into a machine as the Abacus. There are five λ-calculus variables in arithmetic: *a, =, b, +, -* and *c.* Using the λ-calculus, a Church-Turing Machine, as well as an Abacus, performs arithmetic without any baggage or unseen, undetected outside threats. The expression defines the symbols in the framework. This DNA of symbols to abstraction can be engineered for any purpose.

Let that sink in for a moment. There are so many consequences that impact society. They all date back to the abacus and explain why the universal model of computation is also the best solution to computer science in the age of digital convergence. In the digitally converged world of global cyberspace, the software must be as trusted as both the Abacus and the Slide Rule, as powerful as Ada's dream, as infallibly automated as Babbage's hallmark standards and as easy to use as Steve Jobs' Apple Macintosh[88].

Blind trust is unacceptable for the endless, relentless future of civilisation. Centralised operating systems and dictatorial superusers are enslaving society. No branded baggage in a General-Purpose Computer adds tangible value to science, mathematics or society. It only satisfied von Neumann's ambition to be first, and ever since, as the way monopoly suppliers keep a dictator's grip on the market. The branded conventions of General-Purpose Computer Science lock the clients to the supplier. The baggage corrupts the value of flawless results.

Integrity is broken by hackers and malware that remain undetected.

---

88   Mac 1978 Steve Jobs first proposes Apple to develop a next- generation computer.- Wikipedia Article

None of this meets Babbage's standards, while endless *Infallible Automation* remains Ada's unfulfilled dream. The essential characteristic of computer science, the universal model of transparent computation, is missing. Furthermore, because the General-Purpose Computer is opaque, lay citizens cannot understand it. Only skilled specialists cope with the dirty details and obscure settings of hardware and software in the branded concoctions of General-Purpose Computer Science.

One crooked or unskilled administrator, one lousy download, or one clickjack attack contaminates an innocent user's request and immediately escalates into an undetected but fatal attack—international criminals and national spy agencies silently and insidiously prey on life in cyberspace. Undetected and unrestrained, they snoop, spy, corrupt, crash and steal. These criminal acts are unprosecuted because attacks remain undetected. The guilty party is untraced and escapes. It is unacceptable for democracy; it will not satisfy in the 21st century. Both Charles Babbage and Alonzo Church would find this intolerable. It will result in a cyber dictatorship and cap the progress of civilisation.

A UNIVERSAL MODEL OF COMPUTATION

The universal model of computation solves all the troubles and offers far better results. It is the machine of mathematical science used as proven by nature. As a Rose petal is shaped and coloured, it follows the blueprint of a DNA string with substituted variables. The λ-calculus encapsulates and cultivates this natural, private form of computation.

The Church-Turing Machine also grows atomic structures as a species with inherited instincts to survive and the means to evolve gracefully over generations on end. The computational gene in the universal model of computation is Turing's α-machine held safe on the rails of a directional DNA hierarchy. It is a capability-based object-oriented machine code model that supplies the stability and security to evolve incrementally over time.

A patched upgrade and a monolithic recompilation lose the invested value in a complex system. Instead, improvements must progress atomically through the DNA of an application species.

The architecture is atomic. The digital gene is a software survival framework, and the directional DNA hierarchy defines a mathematical machine. A *'single algorithm'* is encapsulated in a context expressed by a few symbols, the λ-calculus variables in a mathematical expression.

The symbols are programs controlled by Church Instructions that navigate the nodal frames as chained DNA strings. The machinery can be trusted because it scientifically executes protected functional expressions by following the DNA blueprint.

TRUSTED COMPUTERS

In 1936, the pre-electronic age ruled, and the work of Church and Turing was not seen then as it later became the two sides of the Church-Turing Thesis. The idea of passing a variable through chained function abstractions, as Ada demonstrated a century ahead of her time, and her dream, explained in her notes, did not resonate. Every expert in the pre-electronic age of WWII believed a handful of computers could satisfy the world, so how they were architected mattered little. Only Alonzo Church and Ada Lovelace saw a vision of universal cyberspace.

Trusted software was unnecessary when the software was in short supply and always homegrown, and computers were isolated. Global cyberspace and the age of micro-electronics made no sense at the end of WWII. Security was an organisational challenge for another four decades as the Cold War raged. Spy agencies ran security; it was not a software issue.

Soon, teams of security-cleared IT professionals carried photo passes for Identity Based Access Control. Identity-based Access control matched the secretive, opaque, and mysterious centralised agencies that ran the Cold War. They locked the General-Purpose Computers in a room or even a basement dungeon, isolated from the rest of the world by guards and approved individuals through Access Control Lists.

Guarded access depends on a name and a password to enter the inner sanctum. These recreations of an Ishtar Gate unlocked access to barricaded sites, sealed buildings, locked rooms, and guarded computer keyboards, but once logged on, a centralised operating system ran cybersecurity. In a virtual machine where all the software actions occur, digital error detection is non- existent.

The pre-electronic age transferred Cold War traditions to the General-Purpose Computer and virtual machines, ignoring software and shared memory. Initially, a one-to-one-to-one-to- one relationship existed between the homegrown Turing machine, a homegrown program, the inventor, and a locked room. While batch processing ruled, the physical form of the computer, the software, and the function all matched.

Everything shifted as software improved, but hardware remained *'backwards compatible.'* Backward compatibility allowed old programs to run on new computers, specifically prioritized by the effort and cost of developing the centralized, superuser, most dangerous software. Backward compatibility increased the life of the hardest to build at the lowest level of software, typically written in C++ and, in the worst case, statically bound machine code[89].

This effort has increased with the growth of complexity in unquestioningly trusted privileged operating systems and all the applications that use these branded services. The branded industry of General-Purpose Computer Science grew out of central memory privileges and stayed this way. The imperative Turing-Commands of the Intel 8080 and all later microprocessors are stuck in the past. At the same time, all-around cyber insecurity grows, attacks get more imaginative and the root cause, von Neumann's centralised, unfairly privileged, shared memory architecture, that ignores science stays unsolved.

Trusted software does not exist. Even the meaning of *'trusted'* is abused. The industry definition is proprietary and vendor- specific. Each vendor has their own branded definition, and *'endorsement keys'* only guarantee that some version of a supplier's software is

---

89   The size of the Windows 11 operating systems is estimated to be between 60 and 100 million statements. Porting this code to another computers obviously takes a huge amount effort.

installed. It does not stop malware, hacking, or a corrupt superuser walking away with the crown jewels. Interworking between *'trusted'* machines still depends on untrustworthy binary data created by others. All interworking is based on blind trust.

Inevitably, this approach, sponsored by an industry consortium, suppresses competition. It is deliberate. Branded conventions and the lack of standards keep a supplier's grip on the market, and the economic impact locks corporate customers to one branded supplier. *Trusted* General-Purpose Computers are nicknamed *Treacherous Computing* by the open software community.

FAITHFUL COMPUTER SCIENCE

When Ada wrote her program for the Analytical Engine, the definition of trusted computers was indistinguishable from mathematically correct. Babbage removed all human errors. Faithful computer science is now a better connotation than over- abused and underachieved claims of trusted General-Purpose Computers. A faithful mathematical software servant on reliable, fault-tolerant hardware systems expresses the essence required by future societies living as virtualised democracies—an infallibly automated mathematical servant, not a capricious, unreliable dictator.

Computer science for the 21st century and beyond must be a faithful servant, equally to all. A level playing field where democracy can both survive and flourish, like a Rose with forms of beauty and forms of petals, in a competitive, hostile world. As a secure servant to society, various democratic solutions will safely coexist in the electronic village. Laws will change, so software security policies must adapt in the endless, evolving future. However, when human conflicts extend into cyberspace, a common denominator is always the same: attacks must be detected, recognised, and resolved immediately on the surface of cyberspace before any digital harm occurs.

By placing cybersecurity into the hands of the users, the foundation stone of democracy is guaranteed. It turns computer science upside down, and the unelected superusers who govern General-Purpose

Computers using unfair privileges will fall. When citizens control their own data and service providers compete by offering the best function abstractions downloaded to a user Namespace, the universal model of computation will flourish.

Users subscribe to the services they need and only share their data as they choose. They own a private Namespace, and nothing leaves without their approval, which is implemented transparently through Capability Keys. The universal model of computation precisely fills the need.

It is how nature creates natural diversity, stability, and generational evolution of life. More than just the computer, it is the software that must be faithful and trusted by cyber society in the 21st century. When software is expressed scientifically using object-oriented machine code at the programmatic level, functions are *'interpreted'* and *'true-to-form'*. The software is framed, calibrated, and faithfully trusted through Capability Limited Addressing and measured MTBF. The Church-Turing Thesis defines Industrial Strength Computer Science based on the universal model of computation. It is the future of computer science.

SOFTWARE SECURITY

Ada's vision was inspired; she was centuries ahead of her time, on the same level as Steve Jobs. She foresaw the endless continuum of *Infallible Automation* as a perfect clockwork computer, faithfully programmed by mathematics, devoid of human errors, assumptions, corruption, and guesswork. She envisioned an infinite, flawless Cyberspace guiding enlightened society to a complete understanding of the secrets of nature. In 1936, Alonzo Church scientifically deduced her vision as a universal model of computation.

However, for decades, the vision was silenced by WWII, and the Cold War echoes. Her vision was bushwhacked by egos, monopolies, and the interest of spy agencies with the flawed claim of 'Homeland Security.' The pre-electronic age architecture of General-Purpose Computer Science cannot achieve Ada's dream or, more seriously, the needs of democracy in a globally shared electronic village. The demands of the 21st century and beyond can only be served by the universal model of computation.

The General-Purpose Computer tilts against the citizen. Ambition, commercial interest, and spy agencies are creating their digital dictatorships. Consequently, society and civilisation are enslaved and exposed to crime and theft. The flawless expectations of mathematics are subverted and diverted by monolithic software, echoes from the Cold War, and a batch- processing history that marked the start of digital computers.

In a short time, Alan Turing worked with Alonzo Church, where he gained his Doctorate and immortalised his name. However, the secrets of λ-calculus and the universal model of computation were ignored in a commercial rush to own the market. As a result, avoidable flaws remain with backwards compatible, branded General-Purpose Computers overrun by the latest software advances, including functional programming and Artificially Intelligent Malware.

Bronowski's coiled spring of nature hides in the λ-calculus and the universal model of computation that shapes and reshapes life's DNA atomically in an evolutionary stable dance. The λ- calculus underpins the survival of active living things, and dynamic software applications are cut from the same cloth. The λ-calculus dictates functional change through a clockwork meta- machine where the coiled springs in atomic frameworks locally regulate life.

The universal model of computation and rules of λ-calculus Variables, Abstractions, Functions, and Applications change the nature of software in a Namespace from an authoritarian master to a faithful, loyal servant, from remote monolithic compilations to on-the-spot, object-capability machine code. Each Namespace is secure, and stabilised software machines survive the tests of time in hostile global cyberspace.

The best a General-Purpose Computer can do is very limited. These concerns are covered in the short paper titled *Capability Myths Demolished* by Miller et al., introduced earlier in Figure 8. However, none of these software alternatives resolve the hardware voids and the default privileges of superusers and remote hacking.

Only the λ-calculus built as the foundation for Alan Turing's discovery of the α-machine offers the universal model of computation to address every point Miller et al. raise. Most significantly, the universal computation model regulates access rights to modular software through evolving security policies. Capability Limited Addressing restricts these access rights through hardware. Transparent lock-and-key security is tangible, and policy stems from the keys in the hands of the users.

Nevertheless, the success of any software alternative in Figure 8 is limited. The unfair privileges of centralised monolithic software of General-Purpose Computer Science distort the binary computer. As Miller points out, statically bound security cannot meet the full dynamic range of software. Compiled, object-oriented, capability-limited software only begins to solve the problems. To meet the needs of global cyberspace, to match Ada's vision, and to achieve Babbage's standard, object-oriented machine code must be built directly into the hardware using Capability Limited Addressing. Object-oriented machine code, when Capability-limited, makes computer science a future-safe, fail-safe science fit for civilizations' eternal future.

Instead of a centralised operating system, a λ-calculus meta-machine must exist. It is how *Infallible Automation* is born, from Babbage's two engines back through the Slide Rule to the Abacus. Only then will security apply across industry suppliers, including networked browsing and downloads from suspect servers. Every web frame, all server browsing, opening emails, reading an instant message, reading attachments, detecting hidden scripts, tracing data marks, and clicking on unknown links that connect to contaminated sites are all secured using this version of the universal model of computation. It is the way of life in a virtualised world. The interactions must be engineered to prevent catastrophe and make every citizen safe. Capability- based, object-oriented machine code is a complete scientific solution.

The von Neumann architecture became a non-starter when the Internet scaled computers as a network. The lightspeed of global communications demands digital integrity be enforced. Industrial Strength Computer Science is the only workable option for the endless future of cyber society. A Church-Turing Machine unites security with functionality through Capability Limited Addressing and object-oriented machine code. Thus, security and functionality are the same. The Babylonian Golden Rule works full-time on the networked surface of object-oriented computer science.

Identity-based access control and software monitors are bound to user identities, and Lampson's ACL security matrix only exists for human users. However, the foundation layer of dynamic software is only regulated by the clockwork of a λ- calculus meta-machine that acts as a built-in security governor bound to the modular functionality by Capability Limited Addressing.

This security mechanism cannot be bypassed and is under dynamic policy control by individual citizens instead of monopolies, spy agencies, and criminals. Babbage described a governor as a *'beautiful contrivance.'* For mechanical engines, governors act in the background as a security mechanism, a lock- and-key to crosscheck critical activities in real time. In a Church- Turing Machine, the Capability Limited Addressing works in the background to cross-check access to every object in every machine instruction. The Babylonian Golden Rule makes functional software as safe as industrial-strength Computer Science. Cybersecurity is the foremost concern of a 21st-century cyber society. Only the best will do.

The best is defined by nature as a universal model of computation and the Church-Turing Thesis. Only a Church-Turing Machine addresses every need.

Instruction by instruction, function by function and abstraction by abstraction, directly under programmed control, the laws of λ-calculus enforce the Babylonian Golden Rule. Capability Limited Addressing administered by individual Capability Keys extends from the surface of cyberspace into the hands of the users as tangible keys.

Implementing the λ-calculus as a clockwork meta-machine at the base of a capability-based technology stack in a Church- Turing Machine guarantees mathematical control over Industrial Strength Computer Science. The meaningful crosschecks govern the modular software—Capability Limited constraints and crosschecking access rights of each instruction. Even a tiny discrepancy or deviation is detected as the first sign of an attack. Aborting the Thread is the immediate, fail-safe reaction.

The safety net is automatic, built into Turing's binary computer by the λ-calculus and Capability Limited Addressing. Automatically, an innate recovery strategy takes over, a regulator in the universal model of computation that, again, in Babbage's words, *'prevents injurious or dangerous consequences.'*

INNATE IMMUNITY

Innate immunity is the hallmark of Industrial Strength Computer Science. An Industrial-Strength Computer protects the mathematics and the digital integrity of a computer. Capability Limited Addressing secures capability-based, object-oriented machine code as the λ-calculus foundation of a Church-Turing Machine. The symbols of mathematical expressions are functionally secure digital objects under lock-and-key control.

A Namespace is the *'strong room'* of personal computer science. A private reading room accessed by Immutable Capability Keys that all belong and are only known to the Namespace owner. The owner issues the keys to the reading room as items in a library, building a directional DNA hierarchy on a need-to-know and with the least-authority basis to meet their needs.

For example, *Read-Only* access approved for a λ-calculus Variable that stores personal data brings the approved reader as a functional service into the reading room. The confidential data need never leave. A fail-safe trigger of the λ-calculus meta- machine crosschecks and limits the instructions that access this Capability Key. The Namespace is a private desktop that belongs to the user that cannot be bypassed. Any data leaving the Namespace is transparently delegated, and it can be scrubbed and watermarked to prevent identity theft, false news, and deep fakes.

The directional DNA hierarchy is a multi-dimensional graph of Capability Keys formed by the assembly of authorised actions. The relationships follow the security mantra of type-controlled need-to-know and least-authority—the same security strategy used to access a strongbox in the natural world. The DNA of immutable Capability Keys encodes the digital Namespace's functional chemistry as the private functional extensions of individuals in cyberspace. The defined mathematics of a λ- calculus namespace is written on a virtual chalkboard. Freedom, equality, and justice that all can appreciate from the earliest days at school.

The object-oriented machine code as function abstractions resolves security using formal set theory. On behalf of individuals, object-oriented machine code uses the universal computation model to separate privacy and functionality. The DNA blueprint structures form to function at every step. The regulated, fail-safe algebra is interpreted and regulated on use. Thus, the raw digital media is protected logically, functionally, and physically in atomic forms as mathematically engineered frames of computation. Its scientifically complete structure is executed in trusted chains of secured functionality, all driven by unique events as variables controlled and determined by the citizen users in cyberspace.

This architecture for innate software immunity is inherited from the Church-Turing Machine, the universal model of computation, the λ-calculus, and Capability Limited Addressing. It improves hardware, reduces the software code base and increases software quality. The instinctive, nervous reaction to errors takes place on contact. Any violation is a signal from a Babbage *regulator* to catch digital infections and corruption *'red- handed.'* It mimics the Darwinian rules of survival and adaptation as an evolutionary, stable species with strength, diversity, stability, and security.

The graph of Capability Keys encodes the security through the functional composition of a programmed application. Enforcing the Babylonian Golden Rule safeguards infallible mathematical automation. There is no centralised system administrator when the software is defined mathematically. The object-oriented machine code forms the mathematical chemistry as a Namespace.

The intimate, point-to-point working relationships are linked as object-oriented machine code, regulated by delegated Capability Keys. As the DNA evolves, the blueprint still prevents wild, unlawful software. The functional organs can replace any damaged tissue. Point-to-point connections and purposeful Capability Keys are power levers that control a computational species's software limbs. These digital limbs work in parallel and safely across networked cyberspace. While a transparent Capability Key has network reach, the scoped security policy is always local, immediate, and under programmed control to avoid unfair privileges.

UNFAIR PRIVILEGES

The test for Industrial Strength Computer Science is to import or paste malware into a software application and measure the time the computer takes to detect it. The quicker, the better. However, with General-Purpose Computer Science, it can take years to uncover some crimes. Detection depends on boundary checks. Boundaries are ever more fragmented in the monolithic compilations of General-Purpose Computer Science. The universal hack of a virtual machine is to hide malware in a tainted download. Frequently, this attack is disguised as a click, highjacked by a forgery, found by a phishing email or a link in an instant message. Digital fraud is frequently buried in a browser page. Some of these attacks are never discovered, others take years to find, and the press publishes a few. Some are only disclosed by a human tragedy. A Church-Turing Machine like PP250 measures malware discovery in single-digit instruction counts on contact in micro-seconds. Malware, hacking, bugs, and accidents are all caught red-handed on the spot at the time, but before a digital mistake.

Hacking spreads through remote access, and malware spreads by downloads. Accidents spread by cut and paste and other replication mechanisms. Hacking uses unfair privileges, while malware spreads through instruction privileges. Identity-Based Access Control is hacked using stolen credentials from a massive data breach, a spoofing email or a confused deputy attack. The lists include active worksheets and documents anywhere malware can hide[90].

---

90   Using copy-and-paste is an 'untyped' action and the results cannot be blindly

Binary software monitors cannot catch the infinite worst-case malware problems in a virtual machine. In the time it takes to pass through airport security, an enemy state can plant malware from a thumb drive on a laptop. Unfair binary privileges within and between virtual machines ignore the assumptions of blind trust. It was the first mistake after WWII, later amplified by virtual memory, superusers, central operating systems, timesharing and Identity-Based Access Control during the Cold War. Malware and hacking succeed and remain undetected because blind trust and unfair privileges pervade General-Purpose Computers. The essence of malware and hacking is the misuse of exposed privileges created by a lack of mathematical science.

Malware in a virtual machine can wipe the memory clean with a single general-purpose instruction. A superuser attack can load a new operating system. The Cold War paradigm of Identity Based Access Control fell apart when the internet introduced dynamic software and downloads. These shared security mistakes must now be relearned by every generation of software programmers, over and over, again and again. As taught today, computer science comes nowhere close to creating the quality results needed to survive endlessly in hostile cyberspace.

In a Church-Turing Machine, software inherits cybersecurity through inbuilt *Infallible Automation*. For example, a copy-and-paste in a Church-Turing Machine includes the DNA strings through capability-limited meta-data that protect the woven digital material. Its cut-and-paste operation is pure functional science without any malware or hacking. The DNA string stays in place because the form matches the capability-limited object- oriented machine code function. The same security exists when browsing, for download or email, for False News and Deep Fakes.

The binary computer limits virtual reality, while the λ-calculus meta-machine makes virtual reality work safely and correctly for time everlasting. The digital implementation is not abstract when the

---

trusted. Even without interfering clipboard malware the action can fail to paste current data perhaps using some previous copy. It is a risk when dealing with private data, such as pasting password, into a data field that are public, perhaps on Facebook or Twitter.

abstractions are hardened. Because Capability Limited Addressing governs functionality, using the Babylonian Golden Rule, it also regulates security. Error detection now takes place at every level of software abstraction. The function abstractions are hardened as digital components with a calibrated MTBF. The object-oriented machine code of mathematical types forms the gene pool of protected digital organs as a mathematical species. Each encapsulated part carries a DNA string that shields all operations on the digital frontier. It includes cut and paste and any form of replication. Everything is type-controlled, with a secure atomic structure. Malware infections and hacking are locked out of contact. The encapsulated machine types all obey the Babylonian Golden Rule.

ALGEBRAIC COMPUTATIONS

Building software as the symbols adds a formal algebra that reaches across cyberspace. The object-oriented machine code in a DNA hierarchy provides a security graph that matches the functional structure. It is a limited graph of functions and permissions where the Capability Keys isolate outside concerns. Dynamic binding frames the individual functions and abstract organs that create the digital senses, such as remote limbs, complex eyes, ears, hands, and feet of a tangible (capability- based) software species.

The species is designed to solve problems and is focused on the objective, just like telecommunications for PP250. The design is not distracted from the task by cybersecurity concerns or any branded baggage. Security matches functionality as a local concern. The algebra is a limited context defined as a frame of λ- calculus Variables defined by a nuclear C-List.

The algebra scales without global limits while controlling every user's confidential data. The DNA can differ, but the atomic computation rules following the λ-calculus always stay the same. The nature of a Church-Turing Machine automatically secures the digital borders of software. Shaping the functions takes place through two complementary levels of programming. One is mathematical and functional, and the other is material and procedural. One is the construction of a function abstraction, and the other is the definition of DNA for Namespace applications.

These two levels of programming separate the implementation of independent concerns. Development and deployment are simplified by replication, modularity, substitution, and hiding. The beauty of a woven mathematical cloth forms cyberspace as mathematical function abstractions, exactly as Ada first envisioned programming.

INDUSTRIAL PROGRESS

The universal model of computation is nature's solution to stable, steady improvements in hostile cyberspace. Infallible automation is vital to prevent internationally inspired-human hostility. The clockwork solution to software survival is always the same: inherited from the Church-Turing Machine.

Functionality, stability and reliability improve generation after generation by extending the DNA without changing the form or functions of existing abstractions, and the application can improve by improving individual functions in selected abstractions. The invested value accumulates because the software survives. To ignore λ-calculus only invites corruption and crime into computer science— the λ-calculus seals mathematical interactions in the universal computational model. Object-oriented machine code is media-tight, function-tight, and data-tight at the dead centre of the Church-Turing Thesis.

It materialises Ada's dream with Babbage's *Infallible Automation*. A Church-Turing Machine leaves no unguarded space or special privileges to misuse in a weaponised attack and a lost cyberwar.

Ada foresaw the progress of science in chemistry, biology, and physics as linked to computer science by the forces of Mother Nature. Nuclear science defines all progress in the age of electronics. It started with splitting the atom and ending WWII, followed by the invention of the point-contact transistor in 1947 and 1953, the discovery of the double helix, the twisted-ladder structure of deoxyribonucleic acid (DNA) and understanding of molecular biology and the stable evolutionary genes of Darwinian evolution.

The tragedy of the General-Purpose Computer is that the λ-calculus discovered in 1936 was ignored. It could be forgiven when WWII ended, but in 1977, the case was clear. Now, four to five decades later, time is running out.

Still, the leaders in the industry ignore science in the pursuit of self-serving ambition. Cybersecurity is swept under the carpet, still encouraged by government spy agencies who direct research budgets and ignore the needs of virtualised democracy. The atomic deconstruction of monolithic software is the key to the future. Only dictators will win using virtual machines, compilers, privileged operating systems, and superusers.

For less hardware cost and far less software effort, all the baggage of General-Purpose Computer Science must be ditched and replaced by a Church-Turing Machine and Industrial Strength Computer Science. It levels the playing field by using nature's universal model of computation.

Removing blind trust and default rings of privilege brings pure mathematical science to the computer industry. The Church-Turing Thesis leans forward to offer Industrial Strength Cyberspace to automate all human practices, removing unfair privileges and levelling the playing field of computer science.

Then, Cyberspace becomes a global platform immune to hacking and void of hidden, undetected corruption. A public utility that is the servant of individuals, society, nations, and humanity Industrial Strength Computer Science is the mantra for a future-safe Cyber society. Any bugs are detected on the first occurrence, and systems recover without a service breakdown, an urgent patch, a destructive recompilation, and an all too- complex regression test. The endless cost of ownership from zero-day[91] attacks and monthly upgrades disappear.

---

91   A zero-day exploit is a cyber-attack that exposes an undiscovered weakness in

The advantage is dramatic. Development takes months instead of years, weeks instead of months, and upgrades are distributed atomically and automatically. The half-life increases from weeks to decades, if not centuries, dramatically reducing the cost of ownership. Qualified service levels are continuously sustained at the highest levels without needing a patch or experiencing loss, corruption, or theft.

All this is achieved through modular software with calibrated MTBF of the object-oriented machine code. Knowing the MTBF of each function abstraction promotes graceful evolution, functional stability, and reliable service. Maintenance is simplified to improving the weakest link, self-identified by the worst MTBF. This faithful formula, started by Babbage, is trusted because science reigns and human interference is prevented. Infallible automation in a Church-Turing Machine guarantees success. Hacking, malware, blackouts, and urgent patched upgrades are all purged. A dictatorial takeover by Artificially

Intelligent Weaponised Malware is blocked, while data privacy and democracy remain in the hands of citizens.

---

software that is exploited before a fix is available from the supplier.

CHAPTER 10

The Mathematical DNA

"The purpose of abstraction is not to be vague, but to create a new semantic level in which one can be absolutely precise."
Edger Dijkstra, 1930-2002, was a Dutch computer scientist who influenced software as a discipline from practical and theoretical perspectives.

Both the Abacus and the Babylonian Golden Rule are products of natural selection. Natural selection wins by out-surviving and out-evolving all the competition. The speed of evolution prevails over the tests of time as the essential mechanism that finds winners over losers. Digital computers and software are no different; Mother Nature governs everything. The Abacus and the Slide Rule survive because science and society are served equally and faithfully.

Survival and evolution are questions of efficiency over the competition. The simplest form with the best function will maximise performance and minimise loss. The future-safe survival of computer science is the same. By exactly matching digital boundaries to functional symbols grouped as a class, the essentials replace the accidental, and a Church-Turing Machine will replace the General-Purpose Computer.

As Dijkstra understood, enhanced semantics emerge from assembling essential functions, removing the accidental characteristics to abstract a task as functions. As the Abacus forever replaced a general-purpose pile of stones, the synergy of digital form to programmed function as a Church-Turing Machine delivers more functionality and improved quality for less effort with improved results.

For example, a human hand frames the abstraction of a rail in the Abacus. Each rail as an assembly is a unit of abstraction. It is an atomic machine of hands. The Slide Rule further embeds several functions, side-by-side, embedded, pre-programmed algorithms that work towards Babbage's trademark of *Infallible Automation*.

Interestingly, the Abacus and the Slide Rule use replication to scale and grow, but in different ways. The Abacus scales by adding rails for larger numbers. The Slide Rule scales by adding new algorithms on the same sliding rods. One scales the size, and the other scales the functions. Atomically, the λ-calculus supports both replication and substitution. As λ-calculus symbols, the computational concepts of Variables, Abstractions, Functions, and Applications in a Namespace frame the universal model of computation.

In a Church-Turing Machine, the universal model of computation abstracts the λ-calculus variables to the simplest functional forms for easy replication to construct a higher-level abstraction from smaller ones. Its scalable process can continue indefinitely. Mathematics remains pure and straightforward recursive steps, while transparent security and functionality are enhanced. The primary aim for each abstraction unit is the fundamental expression of ideas without any unnecessary or accidental baggage, as the Abacus and the Slide Rule demonstrate so well.

Computer science was a problematic hardware challenge after WWII and throughout the Cold War in the pre-electronic age that ended with the Intel 8080 microprocessor. Any computer was challenging to design, and networking was over the horizon. These early computers could ignore λ-calculus. As applications grew, memory limitations led to virtual memory as the accepted solution for growth. It was an unfortunate mistake to abstract large storage systems as linear virtual memory before abstracting programs as mathematical functions using the λ- calculus.

Prematurely, computers competed over a physical memory, peculiar machine instructions, memory word length, and all the accidental baggage and default privileges that defined General-Purpose Computer Science. None of this opaque nonsense must be mastered at school to compute the mathematics written on the chalkboard. None of the baggage invented by General- Purpose Computer Science is needed.

THE ART OF ABSTRACTION

Instead of abstracting physical memory into pages, a Church-Turing Machine adds a handful of Church Instructions to abstract procedural programs into purely dedicated function abstractions. Function abstraction like the one Ada Lovelace wrote leads to software machines that mimic the Abacus, the Slide Rule, and Babbage's flawless mechanical engines. The alternative uses the λ-calculus to define function abstractions as scientific instruments. Scientific instrumentation reduces the software to the needed imported functions and data variables kept private by the universal model of computation.

A function abstraction in a Church-Turing Machine is where calculations occur in chains, following the sequential expressions teachers and scientists write on a chalkboard. Each abstraction is a machine that implements a function or a set of related tasks as a class. As such, there are easily checked limits and simple methods of use; for example, the abstraction of a decimal digit as a hand is unconfused by numbers more than nine or less than zero.

The abstraction of a decimal number survives because it intrinsically stands for an atomic digit rather than attempting to represent some general form of a larger number using a pile of stones. Over millennia, the faithful Abacus has served as a trusted servant. The Abacus remained a practical way to sell and buy goods throughout time. It can still be found at work in village markets around the world. The abstract computations are dynamically bound together using a wooden array of beaded rails that physically keep the individual calculations organized most effectively and the overall result on track.

As with the Abacus, the Slide Rule has a mechanical structure as a machine chosen to make the task easy, reliable, and faithful to nature's science and society's needs. The atomic machinery is shaped to express a function as a reliable clockwork mechanical task. Mathematically speaking, the structure is enforced by the built-in DNA hierarchy of the machine as type-specific engineered actions. The easy actions work for all, independent of their skill or level of education.

In engineering terms, the structure follows a directional DNA hierarchy of mathematical types built as a dynamic machine. The DNA defines the mechanics of the machine as a species. The DNA of the Abacus is an array of decimal digits. Symbols represent the relationships as a mechanical representation of the mathematical expression that underpins the structure. Each rail is a digit in the power of tens collection, and digits are named accordingly. The rails all work the same way, using the same replicated structure, and only the collection as an abstraction of a large decimal number needs to understand the roles of the individual rails. The machine computes the mathematical statement of symbolic logic with substituted λ-calculus variables. The digits and the chosen function, add or subtract, are substituted as required. The framework is specific to the case. It is not general-purpose; it is a copy of the process taught at school that only works for decimal addition or subtraction.

A pile of stones is a general-purpose implementation, but it cannot sustain reliable results. Furthermore, the algorithm is a peculiar proprietary solution that can be performed differently. It is extremely slow or opaquely complicated and far more challenging to use and share with confidence. Natural selection rejected and refused to use these opaque algorithms and peculiar conventions with no memorable justification or guard rails to detect hidden errors. The monolithic pile cannot scale incrementally through replication and requires complex, easily forgotten best practices.

Any misunderstandings or missed operations lead to an undiscovered error that causes conflicts between users. In global cyberspace, general-purpose computers lead to cyberwars that can only be resolved by traditional wars. On the other hand, the abstractions in a λ-calculus solution offer a scientific framework of symbols to solve all these problems created by uncertainty.

Indeed, the disputes with Fake News and Deep Fakes that subvert democratic society can all be reduced by a Church-Turing Machine using the λ-calculus and capability-limited addressing.

CHAINED ABSTRACTIONS

Intrinsic abstractions like the replicated rails in an Abacus and replicated algorithms in a Slide Rule are atomic. Indeed, the objective is to reduce complexity into simple atomic components where the mathematics is pure and without branded conventions; as such, a mathematical abstraction is future-safe and survives forever. The multi-thousand-year history of the Abacus, the Slide Rule's multi-functional power, and Ada's function abstraction for Bernoulli Numbers well demonstrate this irresistible power.

These nuclear machines have engineered guard rails as safety regulators. The safety mechanism is the DNA that defines the mechanical framework of each mathematical machine type. The DNA grows as each abstraction is constructed as an extension of atomic ideas. At each functional level, the extended machine types are clearly defined as functional types using the exact DNA mechanisms that implement the nuclear components.

The Array, as an abstraction in an Abacus, is not a monolithic decimal number but an array of individual decimal digits. The distinction is essential. The algorithm for the collection as a framework of decimal rails only deals with ordering overflow or underflow events. The array algorithm cycles through the individual decimal numbers, performing the simplest form of addition and subtraction. The event-driven function of the array is specialised to resolve the orderly mechanism for a carry and borrow.

The array has a different algorithm from the decimal addition and subtraction functions for the individual rails. The array algorithm only involves exception events generated by the decimal level of computation. Further, and again importantly, when the exception is passed as a λ-calculus variable, the exception handling is also abstracted to the purest form. What is more, the architecture using the universal model of computation allows parallel computations to occur. Its flexibility is vital for any global application that demands a distributed implementation.

In every case, abstractions hide these details. They are unnecessary for others to know. Abstractions are the core mechanism for implementing the 'need to know security rule.' The higher levels of abstractions are functional organs bound to private implementation details by the namespace DNA. The binding relationship is the list of symbols the algorithms need, but the implementation details remain hidden by the links in a chain of abstractions built from abstractions.

In a Church-Turing Machine like PP250, a Capability Key linking two DNA chain abstractions uses the novel {Enter.someAbstraction} access right. The permission is only granted to approach the Ishtar Gate of the walled and defended abstraction. Direct access is not allowed. The C-List that stores the DNA string is a hidden structure. A visitor approaching the gate, as it were on foot, cannot see the list of Capability Keys that frame Babylon's private, defended functions.

A private C-List is the walled and gated nodal list defining the hidden workings of an *Enter Only* software Black-box. Each Black-box is a fail-safe function abstraction with the physical property of a calibrated MTBF. The specialised algorithms compute the mathematical result, calling on other *Enter Only* Black-box abstractions to simplify the work. Thus, the DNA implements the relationship between protected abstractions. An example structure is shown in Figure 21 for a private digital copy of the Abacus.

The Capability Keys form the directional DNA hierarchy of C-Lists, as shown in Figure 21, where EO stands for '*EnterOnly*' access right. The immutable Capability Keys in the C-List defined the digitally chained relationships that are transparently bound on use.

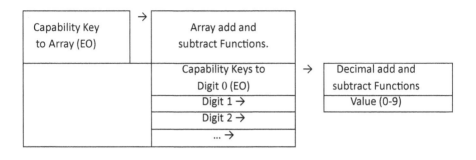

FIGURE 21: A DIRECTIONAL DNA CHAIN OF ENTERONLY (EO) CAPABILITY KEYS FOR AN ABACUS

By example only, the functions for a decimal number are simplified to single digits and a flow event, (+1) for overflow, a negative borrow (-1) or (0) for no overflow. These events are ordered for resolution by the array methods. Several solutions are available to pass the event to the next digit in the array. Besides the synchronous approach, other implementations exist for distributed systems or heavyweight computations where parallel Threads are needed and can improve performance. The Capability Keys define these options and do not change the machinery or require a centralised operating system.

The directional DNA hierarchy secures the overall functionality of a computation in a network built on a simple but solid atomic foundation. The nuclear machines and the array remain intrinsically pure; there is no unwanted baggage or need for a centralised dictator.

The essential functions form as chains by the DNA serve the same purpose, either in wood or as Capability Keys. The machinery is engineered to prevent errors. With the generic properties of Capability Keys software components held in place, they stay true to form. The DNA is built atomically from inside to out, protected from errors at every step. When form matches function at every level of the directional DNA hierarchy, no room exists for undetected errors or outside interference.

This atomic abstraction is a life form's survival and evolution mechanism. Unfortunately, this machinery for abstraction is missing in the shared memory of a General-Purpose Computer. The memory system is a pile of bits without any atomic fabric for protection. The only framework is the physical, monolithic compilation for batch processing where form no longer matches function. The monolithic form is limited to batch processing. Anything else is *'home-grown'* from design, maintenance, and security perspectives. Worse still, monolithic software cannot scale into an open network.

When Ada Lovelace wrote the first software program, she was empowered because she was unhindered by incoherent baggage. She only needed her intrinsically pure mathematical skills and Babbage's core mechanical function abstractions. The mathematical statements she learned when studying mathematics at university applied directly to Babbage's Analytic Engine.

THE DNA OF COMPUTERS

B abbage abandoned his unfinished Difference Engine for his Analytical Engine because he discovered flawless automation through the mechanical abstraction of intrinsic mathematics. As a clockwork machine, the Analytical Engine prevents all human errors. It flawlessly computes using the pure simplicity of mathematical symbolism as learned at school and university. He included the ability to store numbers and print the tabular results to Ada's chosen degree of accuracy without needing a human print setter. Like the Abacus and the Slide Rule, the Analytical Engine is built from dynamically bound mathematical abstractions.

The programs, including Ada's example, are extended machine types. Function abstractions that extend the machine functions with new functions that obey the universal model of computation. For Babbage, the atomic machinery starts with the function abstraction of decimal numbers. So, the Analytical Engine combined the Slide Rule and the Abacus, which Babbage intended to be powered flawlessly by steam.

His machine's cogs and levers are guarded mechanical frameworks engineered to prevent mechanical and human interference. In groups, they offer mathematical functions limited by the physics of nature. All this is to avoid errors misuse, and maintain mechanical integrity. In a Church-Turing Machine, the engineered abstractions are walled and protected by a private Ishtar Gate using Capability Limited Addressing. Capability Keys check any credentials and lockout spies and criminals. The Babylonian Golden Rule is simple, dynamically enforced boundaries that detect errors and prevent damage.

Now, this idea can be replicated atomically for the essential core of functions in mathematics, and the analytical engine can be replicated digitally. A machine for mathematics to calculate and print perfect results, infallibly automated by removing human practices. Add the inherited instincts of λ-calculus, and Turing's α-machine becomes the digital gene in a Church-Turing Machine. In a Church-Turing Machine, an abstraction creates a digital frame of functional software computed in a chain of DNA using the universal model of computation. The universal model of computation separates the data from the atomic functions. The frames are functionally reusable by parallel computations in separate Threads with privately substituted data variables. Each thread performs the calculations for one set of (event) variables while added function abstractions (including Ada's) extend the robust nature of the machine instructions. Adding new mathematical functions does not change the nature of the Thread-driven computer. Replication from a DNA blueprint is crucial to dynamic survival through reproduction and recovery.

Replication and reproduction are elegant, functional mechanisms in themselves. Increasing the size of an array breaks no rules and introduces no new concerns to upset the balance of the design. Extension by replicating an existing function adds another pass to an existing loop. Another rail in a digital Abacus is just a dynamic copy of the object in the DNA string. The method of the Abacus as an array performs an extra recursive step to carry the additional result to the more significant copy of a digit and create a larger number.

The cascaded relationship between atomic components is a phenomenon of the DNA and extended machine types. It does not depend on physical factors such as memory size or network location; Capability Limited Addressing solves these problems. A monolithic compiler and an operating system hinder reasonable solutions to the real-time issues of dynamic worldly activity. Scaling is simplified when applications only depend on mathematics and scientifically associating simple methods of each abstraction with instance variables carried by Threads. All that is required is an Object-Capability Macro Assembler. The DNA structures named relationships as symbols in orderly arrangements of mathematics, functionally secured by intrinsic methods of use. In each case, a leaf node in the DNA tree can grow and die during the mathematical progress of the computations. It is an essential living process of change, growth, and evolution common to all life-supporting software systems, including the global telecommunications application that sponsored the PP250. It is vital to the virtualization of every 21st- century software application.

Any change significantly impacts a pre-compiled and shared binary image that is all statically bound together. There is no atomic blueprint for protecting the dynamic change without harming a shared image. Over time, as things evolve and change, wear and tear take place. The framework of a General-Purpose Computer is just a pile of binary stones without a form to the functions. A pile of rocks gets kicked around, some are lost or stolen, and problems emerge. No different from Babylonian traders when they chose the Abacus over a general-purpose pile of stones.

The λ-calculus is like the Abacus from so long ago; it solves the same problems of mathematical structure. A meta-machine adds this Industrial Strength to Computer Science. It starts in the basement, where the meta-machine structures the universal model of computation using object-oriented machine code. Its platform is mathematically correct, scientifically fair, functionally safe, and interference-free. The foundation technology begins with the laws of λ-calculus instead of starting with an overstretched Turing machine and von Neumann's shared heap of memory.

The universal model of computation shapes a functional execution to match the digital form of the software design. The boundaries of the Babylonian Golden Rule are then enforced using Capability Limited Addressing.

On-demand, the directional DNA hierarchy enforces digital boundaries to match the functional expressions. The expressions are chained together in Threads that work in parallel as transactions serving events as requested by the natural or the digital world. Events that serve society and reach into every corner of an electronic village that demands *Infallible Automation*.

Composed by mathematics, the DNA contextualises individual frames of computation with digital boundaries and digital access rights as the approved links in chained expressions with variable relationships. These hardened Threads can reliably use extended machine types for life-supporting applications. Capability Limited Addressing protects the software and is bound together as functional digital media by the λ-calculus meta-machine.

IMPLEMENTATION HIDING

Reconsider the twenty-five steps in Ada's first computer program. It is not a program that could run on a General-Purpose Computer. However, it is a function abstraction that could run, generations later, on a Church-Turing Machine in a mathematical namespace. The mathematical DNA for the computation is clear; it is expressed by a few symbols and expressions on a mathematician's chalkboard. Babbage assembled these mathematical objects, and in the implementation, the machine code of cogs and leavers were all hidden behind a functional interface. The symbols she wrote on the chalkboard included a loop shown in Figure 23, taken from her Note G. The symbols have relationships in expressions, which define access rights in the ordered sequence of calculations. The calculations are chained together as a scientific flow with a DNA of symbols and expressions. The expressions have nodal significance taken from the chalkboard.

When the chalkboard is realised this way, *Infallible Automation* results. The universal model of computation substitutes values for each protected symbol enforced by the directional DNA hierarchy. The Capability Keys in a Church-Turing Machine serve this purpose. Some threads run in sequence, others as parallel calculations, some as local, and others as remote. In a Church-Turing Machine, these relationships are vital to global cyberspace. They are ignored and cannot be reliably implemented in a statically bound General-Purpose Computer.

Using nature's universal model of computation simplifies and secures global cyberspace. Nature's success is safely modelled and computed by the digital world. A λ-calculus meta-machine is the foundation layer of the new technology stack for computer science. Frame by frame, the meta-machine logically and functionally encapsulates distributed object-oriented machine code. The machine code is bound to objects by the context registers of a Church-Turing Machine.

Turing's imperative commands that manipulate binary data equate to Babbage's cogs and leavers, but they remain exposed. In a General-Purpose Computer, this code is statically bound to physical memory that cannot be hidden from outside attacks. The variables are wired to fixed storage locations accessible to all code in a virtual machine.

In a Church-Turing Machine, variables are dynamically bound, hidden by the functional interface of Capability Limited Addressing and can be reused just as Babbage assembled his cogs, wheels, and leavers as dynamically bound functions. The object-oriented machine code is hidden by the functional interface of a generic Church Instruction that dynamically binds the substituted variables as a class, a function, and the required parameters of the call.

Babbage designed this hidden assembly level of the Analytical Engine. Translated by Ada from L. F. Menabrea's memoir, he began with the *'four operations of simple arithmetic upon any numbers whatever.'* Said another way, the lifetime of the mathematical primitives built then from mechanical parts or today as object-oriented machine code is future-safe. It will last if the hardware allows without any need for change. Moreover, the program can be easily ported to the new hardware as technology improves. Costs shrink as branded skills evaporate, and porting costs disappear as the age of everlasting industrial strength computer science takes over.

Ada Lovelace did not program these hidden details; they were assembled just once by Babbage and his engineers as they built the engine. Using Capability Keys and object-oriented machine code, the same is true for a Church-Turing Machine. The manufacturers could include the same core mathematical functions created by Babbage. These core functions define the essence of mathematics, which deconstructs monolithic software into functional digital components as a Namespace. Competition exists between alternative suppliers based on price, performance, security, and simplicity of use.

Another example is the calculation of Bernoulli numbers, programmed by Ada in 1845, as the variable 'B' in the generic expression of symbols shown in Figure 22 and for Ada's example, as found in her Note G, in Figure 23. The program she published in her tutorial evaluating and explaining the flawless power of Babbage's 'Analytical Engine' took just twenty-five steps, and instructions included a loop. The loop of thirteen steps is shown in Figure 23. She documented the entire program in Note G of the Appendix on the Analytical Engine[92] on one sheet of paper.

$$\frac{x}{e^x - 1} \; - \; \sum_{n=0}^{\infty} \frac{B_n x^n}{n!}$$

FIGURE 22: THE EXPRESSION OF BERNOULLI NUMBERS (B)

---

92   See https://www.fourmilab.ch/babbage/figures/menat6_1-5k.png

The Analytical Engine and the Church-Turing Machine descend from the Abacus and the Slide Rule. They preserve the Babylonian Golden Rule of form matching function as a protected boundary framework. For a Church-Turing Machine, the structure of Capability Keys determines the order of resolution of the symbols as a directional DNA string of Capability Keys. On the other hand, a General-Purpose Computer is like a pile of stones endlessly attacked by the feet of humanity competing in a shared but conflicted digital marketplace.

The universal model of computation uses the machine language of the Namespace. These pure symbolic expressions hide the often ugly but unavoidably complex machine design. They were hidden as object-oriented machine code for λ-calculus function abstractions behind functional Church Instructions that standardise the code as functional essentials. It was Charles Babbage's goal as he designed his two engines. Ada Lovelace needed to know nothing about the complex mechanical details that Babbage and his team of experts designed and implemented. He hid the details as mathematical functions, which formed the mathematical machine language of his Analytical Engine, as Ada shows in Note G.

Only the (few) core mathematical functions are assembled as the cogs and leavers of *'machine design.'* These core abstractions of the Analytical Engine elevated Ada's program to a purely functional level above the computer's hidden cogs and leavers. The first layer of Babbage's clockwork functions equates to the atomic level of object-oriented machine code of a mathematical namespace of a Church-Turing Machine. These functions assemble the mathematics as modular, fail-safe and future-safe components. They extend the machine instructions as mathematical types or *'extended machine types'* as we call them in the PP250. The extended machine types build the core functions as trustworthy as any hardware. They are usefully grouped into application-specific Namespace sets, starting with the machinery for mathematics.

13	−	$^1V_6 - {}^1V_1$	$^2V_6$	....	$\left\{\begin{array}{l}^1V_6 = {}^2V_6\\ ^1V_1 = {}^1V_1\end{array}\right\}$	$= 2n-1$ ....
14	+	$^1V_1 + {}^1V_7$	$^2V_7$	....	$\left\{\begin{array}{l}^1V_1 = {}^1V_1\\ ^1V_7 = {}^2V_7\end{array}\right\}$	$= 2+1 = 3$ ....
15	÷	$^2V_6 \div {}^2V_7$	$^1V_8$	....	$\left\{\begin{array}{l}^2V_6 = {}^2V_6\\ ^2V_7 = {}^2V_7\end{array}\right\}$	$= \frac{2n-1}{3}$ ....
16	×	$^1V_8 \times {}^3V_{11}$	$^4V_{11}$	....	$\left\{\begin{array}{l}^1V_8 = {}^0V_8\\ ^3V_{11} = {}^4V_{11}\end{array}\right\}$	$= \frac{2n}{2}\cdot\frac{2n-1}{3}$ ....
17	−	$^2V_6 - {}^1V_1$	$^3V_6$	....	$\left\{\begin{array}{l}^2V_6 = {}^3V_6\\ ^1V_1 = {}^1V_1\end{array}\right\}$	$= 2n-2$ ....
18	+	$^1V_1 + {}^2V_7$	$^3V_7$	....	$\left\{\begin{array}{l}^2V_7 = {}^3V_7\\ ^1V_1 = {}^1V_1\end{array}\right\}$	$= 3+1 = 4$ ....
19	÷	$^3V_6 \div {}^3V_7$	$^1V_9$	....	$\left\{\begin{array}{l}^3V_6 = {}^3V_6\\ ^3V_7 = {}^3V_7\end{array}\right\}$	$= \frac{2n-2}{4}$ ....
20	×	$^1V_9 \times {}^4V_{11}$	$^5V_{11}$	....	$\left\{\begin{array}{l}^1V_9 = {}^0V_9\\ ^4V_{11} = {}^5V_{11}\end{array}\right\}$	$= \frac{2n}{2}\cdot\frac{2n-1}{3}\cdot\frac{2n-2}{4} = A_3$ ......
21	×	$^1V_{22} \times {}^5V_{11}$	$^0V_{12}$	....	$\left\{\begin{array}{l}^1V_{22} = {}^1V_{22}\\ ^0V_{12} = {}^2V_{12}\end{array}\right\}$	$= B_3 \cdot \frac{2n}{2}\cdot\frac{2n-1}{3}\cdot\frac{2n-2}{4} = B_3 A_3$
22	+	$^2V_{12} + {}^2V_{13}$	$^3V_{13}$	....	$\left\{\begin{array}{l}^2V_{12} = {}^0V_{12}\\ ^2V_{13} = {}^3V_{13}\end{array}\right\}$	$= A_0 + B_1 A_1 + B_3 A_3$ ....
23	−	$^2V_{10} - {}^1V_1$	$^3V_{10}$	....	$\left\{\begin{array}{l}^2V_{10} = {}^3V_{10}\\ ^1V_1 = {}^1V_1\end{array}\right\}$	$= n - 3(= 1)$ ....

FIGURE 23: NOTE G, THE LOOP OF TEN MATHEMATICAL EXPRESSIONS IN ADA'S PROGRAM

In this process, the PP250 built extended machine types to support services in the universal model of computation. Critical services previously protected by a centralised operating system now run as sub-routine calls in the universal model of computation. Removing unfair hardware privileges and misleading binary conventions, and without any mathematical void, crack or gap, these extended machine types ran in line without rescheduling delay or binary threats between virtual machines.

For example, object-oriented machine code abstractions for memory management execute as a secure in-line frame in a Thread. Likewise, thread creation, scheduling, synchronization, network communication, and even local device drivers were activated by an in-line functional call to a protected domain activated by the same universal model of computation. These generic services built for telecommunication covered the most complex and challenging worst-case conditions that a General- Purpose Computer struggles to handle well.

The Capability Keys for core services followed the need-to- know and least-authority rules. Memory and processing management abstract the essential mechanism for object creation and removal. Scheduling services using Dijkstra flag objects and facilities management for dealing with hardware devices, with standard input and output functions, and networked communications, everything safely coexists within the universal model of computation where parallel threads use the thread-safe lock-and-key controls. Finally, a Namespace recovery application was provided as a generic fail-safe, diagnostic, and restart service.

EXTENDED MACHINE LANGUAGE

The Telecommunications Applications were built upon these essential services, expanding into an application-oriented, extended machine language: an increasingly higher level, functional language for telecommunications defined by the extended function abstractions of the named Capability Keys. The functions link in chains using the extended types and the named Capability Keys. As Dijkstra knew and a PP250 proved, *'The purpose of abstraction is not to be vague, but to create a new semantic level in which one can be absolutely precise.'*

Ada did the same when she published her program in a mathematical form on a single page. The loop is repeated for each Bernoulli number, taken from her documentation,[93] as shown in Figure 23. Using mathematics, anyone can follow her algorithm. It has survived for generations and, once tested, will never change again. The same 25 steps she drafted on a chalkboard remain valid, calculating the Bernoulli series of numbers.

Note that this loop includes functional λ-calculus variables like A0 and B3, created and named dynamically by the computational thread. These are actual λ-calculus variables, not pre-calculated fixed values. These functional variables are dynamically bound and recalculated on use, as expected by the laws of the λ-calculus with the advantages of functional programming languages.

---

93   Ada's program is a diagram of the computation of the Numbers of Bernoulli and cannot be displayed in this document. The diagram can be downloaded from Note G of Ada's translation of L. F. Menabrea's sketch of the Analytical Engine.

Each Church Instruction is a functional expression using names and symbols for substituted variables and abstractions of any complexity and depth that in a distributed system can stretch worldwide as needed for telecommunications using PP250. The named variables are passed and bound at each step. In each situation, counterbalanced forces are at work. One force is a hard-coded function bound to a symbol, and the other is a mathematical DNA that defines the application as a language for a species.

The dynamically bound λ-calculus variables define a specific species of engineered functionality. Babbage encoded the complex machine code, while Ada mapped the function abstractions into her Bernoulli species. The mathematical equation she chose to evaluate her chosen variables.

Her program loop from step 13 to step 23 is shown in Figure 23. It cycles through the numbers *n* from *zero* to some *max* value. Each step is programmed as clear statements of mathematical logic, such as *2n*, *2n-1*, *2n+1*, and creating λ-calculus variables as functions for use in the loop, *A0 = -1/2 *[(2n-1)/(2n+1)]*. Thus, *A0* must be reevaluated every time for a current value of '*n*'. It is not a static (binary) location or value; instead, it is a function.

Ada programmed the Analytical Engine using punched cards; each card defined one expression of pure mathematical logic. The Bernoulli results were to be saved as impressions on thin copper sheets so that a page of results could be printed flawlessly as a table, without any human involvement beyond the mathematical steps found on her chalkboard, documented in Note G and shown in Figure 23. All errors are removed by Babbage's Infallible Automation process, which eliminates all human practices, including the typesetter. Although he had never built an analytical engine, he proved all the required mechanics in his earlier Difference Engine.

Implementing hiding simplifies all the individual tasks while elevating the '*semantic level in which one can be absolutely precise.*' Monolithic complexity is broken down into functional parts;

each functional part is intrinsically pure and straightforward but semantically potent and fully protected. Function by function, the extended machine types grow by hiding their mathematical details as a sheltered function abstraction.

Implementation hiding is common to both the Analytical Engine and a Church-Turing Machine. Only one engineer needs to know how an individual module as an extended machine type is implemented, and once the mathematical functions are tested, they continue to work as expected. These individual assemblies keep their secrets private from the extended machine. Its ability to hide unnecessary details is significant for cybersecurity, but functional simplification is absent in a General-Purpose Computer.

However good a patched upgrade might be, it cannot cure the problem of digital exposure to foreign downloads. Therefore, the universal model of computation in a Church-Turing Machine also applies to functions downloaded from the internet. A Church-Turing Machine easily achieves this since all software is modular, can be added incrementally, and obeys the universal model of computation.

Resolving symptoms while leaving the root cause firmly in place is the worst band-aid. It costs time and money but does not cure the disease. Patching does not serve the long-term purpose of computer science or society. It cannot save the electronic village from a violent end.

Out of necessity, Babbage carefully hid his mechanical details and only exposed the mathematical functions of the machine. A Church-Turing Machine does the same. By hiding the object-oriented assembly code behind the walls of Capability Limited Addressing, the software is both modular and secure. At the same time, the machine power is increased by the functional syntax and semantics of the extended machine types using the names of the Capability Keys to add context.

Implementation hiding is natural to the λ-calculus, and when the hardware stack starts with a λ-calculus meta-machine using the technology of Capability Limited Addressing, the practical results stretch across cyberspace. The underlying architecture of the λ-calculus shines through by separating functions from data and preventing errors in cyberspace, becoming troubling concerns of the 21st century that end in catastrophic international cyberwars.

The implementation details cannot be touched; only the functions can be used as a protected unit that accepts variables and returns fail-safe results. A Church-Turing Machine is a clockwork software machine with the same flawless performance as Babbage's two engines. The extended machine language increases functionality and reinforces security while separating data from functionality. The most critical requests can execute without unresolved threat vectors by isolating the functional frames from the computational Thread. All the malware threats that attack and confuse the central operating system are avoided when immutable Capability Keys replace corrupt and fraudulent binary data.

SEPARATING CONCERNS

Only one example of an extreme Church-Turing Machine exists as a reference model. As hybrid computers, neither CAP nor CHERI fits the bill. That is a shame. The patent cover on PP250 has long since expired, and the improved technology would solve the digital convergence of global society. Competition is the only way this claim can be proved. The government must take the lead since the established industry will not take the risk, and the danger is too significant to ignore. Investment is needed. It is a proven approach to solving problems and separating concerns.

The dynamic binding of named variables in global, life- supporting applications cleanly separates and simplifies the programming of complicated, globally interacting computer software. The troubling concerns of digital security are reduced to manageable atomic issues and the local management of digital memory.

Pure scientific results are achieved by separating and protecting the atomic details, function by function. Program developers are isolated from outside concerns. Each component solves a class of related issues engineered to resolve all local, worst-case conditions. The modules are then bound together by one set of standard rules using a trusted λ-calculus meta- machine.

Each concern is resolved independently as a single event for black-box frames of computation. The frames are isolated and insulated from one another while still allowing protected variables to pass using the capability context registers of the Thread. The frames unfold as organised by the Church Instructions and Capability Keys in the nodal C-List that define the framework of one function abstraction within one Thread of computation.

Unlike a simple binary value, Capability Keys as λ-calculus variables simplify and speed complex information exchange without confusion or corruption.

CHAPTER 11

Ada's Endless Software Machine

'Mathematics is the art of giving the same
name to different things.'
Jules Henri Poincaré 1854 – 1912, French
Mathematician, theoretical physicist,
engineer, philosopher, and polymath.

In a Church-Turing Machine, when functions are downloaded from any source on the web, they must first be added, without any error, to the λ-calculus Namespace of the download request. As such, the functions of the namespace grow linguistically.

Each added feature, for example, Ada's dynamically created function $A0 = -1/2*[(2n-1)/(2n+1)]$, accepts a λ-calculus variable(s) *'n.'* In Babbage's Analytical Engine, this function is uncluttered by removing all the General-Purpose Computer's opaque, proprietary baggage. A statically bound virtual machine has no mechanism to express such a statement.

Understanding a λ-calculus variable in a General-purpose computer requires huge software overheads, full of threats, risks, and mistakes, including a compiler, virtual memory, an operating system, a programming language, and administrators with the right skill sets.

A Church-Turing Machine easily stores it as one *AO={Enter. function(n)}* capability key as a function abstraction like Ada programmed the Analytical Engine.

Symbols like *AO* mean nothing to General-Purpose Computer machine code because Turing commands are statically bound by a compiler to linear memory. Compile time knows little about run time. Using dynamically defined functional variables requires all the baggage and more of a General-Purpose Computer to solve this simple requirement, which is essential to mathematics. It is as hard a task to anticipate every unknown future malware attack in advance. Both problems are solved by a Church-Turing machine.

However, the five essential concepts of the λ-calculus as Variables, Abstractions, Functions, and Applications in a Namespace are all it takes to define this requirement in a future- safe, fail-safe digital structure using the λ-calculus function abstraction of a Church-Turing machine. A Thread as an implementation of a λ-calculus Application cannot exist in a General-Purpose Computer without all the unquestioningly trusted software baggage to create a virtual machine and more. The difference has profound results.

AN OBJECT-ORIENTED ASSEMBLY

The telecommunication application in a Church-Turing Machine builds up directly on the hardware as the λ-calculus Variables, Abstractions, Functions, and Applications in a telecommunications Namespace. An object-oriented assembler is the only development tool required as an extended machine type built-in as a Namespace used to create new objects. As the application grows, everything stays simple. For example, statements to connect local or international phone calls as a function abstraction in PP250 are as simple as the one shown in Figure 24: A Functional Telephone Connection Request Above with the Assembled Church Instruction, which uses just a Call instruction and some locally available, well-chosen, and well- named Capability Keys.

newConnection = Telephone.Connect (aTelephone,anotherTelephoneNumber) result = Call(class.method,a,b)

FIGURE 24: A FUNCTIONAL TELEPHONE CONNECTION REQUEST ABOVE WITH THE ASSEMBLED CHURCH INSTRUCTION BELOW

Dijkstra would be thrilled; this precise assembler-level statement returns an abstract connection object, with one capability key defining the result. The result is a dynamic construction spanning a world of different telephone systems. The request comprises four well-named

Capability Keys and a single Church Instruction. The '=' sign implies the Church Instruction Call, Change, or Switch, depending on the type of Capability Key, synchronous, asynchronous, or remote. It is all understood and can be assembled from a single macro command.

It is not to deny the hard work it takes to achieve this functionality but to emphasise Dijkstra's point about the power of abstraction, *'to create a new semantic level in which one can be absolutely precise.'* The names of the Capability Keys naturally supply a high-level, self-explanatory syntax. The Capability Keys map to a single Church Instruction using the names to select λ- calculus variables, the Capability Keys within the locally defined context. As in Ada's functional program, full of mathematical symbols, the above functional request directly translates to capability-based, object-oriented machine code. No monolithic compiler or any branded central operating system is needed.

The machine code for such a request names four capability context registers. In this example, the symbolic names for the λ-calculus variables are the namespace Class called *'Telephone,'* the Class function called *'Connect,'* and two telephone numbers to connect *'aTelephone'* and *'anotherTelephoneNumber.'*

The function called 'Connect' is a property of the Class *'Telephone,'* so the PP250 machine code could select the offset value to find the class method indirectly. It is a shortcut that exposes an unknown, untrusted Class to misuse. Since, as in a General-Purpose computer, an attack can forge any binary data and select a random offset. A dangerous private function might be activated. Yet another variation of the *'confused deputy'* attack.

The PP250 offered alternatives for perfect security. The first uses a passive Capability Key to symbolise the chosen class method immutably, thus preventing both fraud and forgery. Now, passive capability keys define both the Class and the function of the Class. A second alternative structures the directional DNA hierarchy to expose only public methods in any *EnterOnly* C-List. The third alternative uses a single guard program to check all credentials, including the function request. The attack vector does not represent a threat to a private interface but is explained to show how external interfaces to foreign code should be handled.

The machine instruction called 'Call' is used to complete this guard function implementation, and then the machine syntax is shown in Figure 25.

$$CR[0] = Call(CR[0], CR[1], CR[2])$$

FIGURE 25: THE MACHINE CODE EXAMPLE

Since the guard offset is fixed, say zero, only three Capability Keys are needed in this evolved case. It is functionally always the same for every class. Without requiring a compiler or any other General-Purpose Computer baggage, the abstractions are self- defined, and the names make them self-explanatory, not in an off-line listing but directly in machine code. These transparent expressions unfold the meaning based on the chosen Capability Key for the selected context.

For example, the instance of *anotherTelephoneNumber* is a potent λ-calculus symbol that includes cybersecurity combined with functionality. The Babylonian Golden Rule implements physical and functional security through Capability Keys, like the keys to cities guarded throughout the history of civilisation. Each key has a name that defines a function materialised in cyberspace as a guarded digital object.

THE HIDDEN SPRING

This self-materialization of meaning, functionality, and security is built into Capability Keys when used as λ-calculus variables. The function abstractions act as nature's hidden springs, unlocking and unfolding endless, infinitely complex or pure and straightforward results in one reusable way. As named abstractions, the Capability Keys represent and hold power in a Church-Turing Machine, not statically bound code, the central operating system or a superuser as in a General-Purpose Computer.

It is exemplified by the decimal number abstractions in the Abacus as an array of decimal digits. The Abacus is an unfolded abstraction of abstractions that scale uniformly ad infinitum. The same machinery in a Church-Turing Machine is black box functional objects realised as functional media, scoped in-depth and detailed as data-tight, function-tight digital abstractions held in place by a directional DNA hierarchy that acts as a skeleton framework. While each step in the evaluation is distinctly private, they are decidedly transparent and fully coordinated as a domesticated digital beast, a servant of the owner, an instance of a guarded software species.

The clockwork laws of λ-calculus govern the transparent service. The rules of the meta-machine encapsulate Turing's α- machine in the typed digital boundaries of Capability Limited Addressing. Capability Limited Addressing encapsulated every dynamically bound action as the Capability Keys are inserted into programmed Church Instructions.

There is no intrusive operating system or any opaque, hidden computational steps; no suspect proprietary conventions or a need to understand linear memory layout. The Capability Keys hide these details. There is no confusion and no fight over the nexus of control, the meaning of variables, the risk of malware, a breakout from the laws of λ-calculus, or an authoritarian administrator who must be unquestioningly trusted.

In a Church-Turing Machine, trust is pervasively framed by provable science. The λ-calculus variables are true-to-form symbols in an algebraic namespace that defines a digital geometry for a flawless mathematical computation. Nothing is opaque, unconsciously shared, or statically exposed, and because everything is transparent and dynamically bound from Capability Keys, scientific clarity abounds, and unfair privileges are removed.

The science of λ-calculus with the technology of Capability Limited Addressing enforces digital media that is always data-tight, function-tight and continuously true-to-form. Furthermore, the digital material interacts directly 'peer to peer' through the authorised names dynamically bound using Capability Keys. The minimalism is highly efficient. A service call to a secure servant frame of computation starts synchronously within a common Thread and, depending on the need, unfolds actions as synchronous, parallel, or remote threads through the nodal C-Lists. Each case is governed by transparent concerns, locally approved authority, and the abstract power of λ-calculus, made physical by Capability keys, with type-limited access rights to tunnelling the process forward within digital guard rails.

There is no centralised superuser to manipulate events, steal data, or compromise the computer with single points of failure — no complex management of shared memory or risky proprietary calls to unguarded but privileged systems. Further, all the endless varieties of 'Confused Deputy' attacks (recall Chapter 4) disappear, along with all the corruptible user privileges. Tricks and forgery that pass fraudulent (binary) conditions to steal secrets are all avoided by enforcing pure scientific logic.

Enforcing the Babylonian Golden Rule on software as data- tight mathematics through expression found on a virtual chalkboard is all that is needed. There is a reason that children learn to compute the algebraic expressions of mathematics symbolically. The reason is reusable simplicity; the directional DNA hierarchy structures the simple frames of computation. The frames encapsulate the digital objects as a fail-safe computation in a Church-Turing Machine, no different from the operations children first learn at school.

Simplicity allows each software part to be checked and measured for reliability. The measured MTBF confirms or denies the Industrial Strength and the quality of the future-safe, cyber- secure software. The highest failure rate has the lowest MTBF in a namespace. It is the candidate for change and evolution; nothing else needs improvement. The software machine is fail- safe and no threat to a life-supporting service. The machine recovers the service, and no permanent damage that harms a General-Purpose Computer occurs.

The Church-Turing Machine exchanges blind trust for mathematical proof; it is an evolutionary, stable, future-safe, and fail-safe implementation that guarantees software runs as a fault-tolerant machine that serves both science and society.

Applications in Church-Turing Machines exist and work atomically, like a multidimensional fishnet or the cells in a living organism. There are no unnatural boundaries with opaque monolithic functionality. The nature of the atoms in a λ-calculus namespace transparently solves the separate problems of computer science cleanly and scientifically.

The result is a stable evolutionary fail-safe computational digital network—secure cyberspace for any size of civilization needed by an electronic village in the 21st century. Combining λ- calculus and Capability addressing, separate functional models resolve all worst-case conditions scientifically. It is a confusion- free, fail-safe design approach to building a complex global software machine, functionally solving each worst-case condition as a uniquely local problem. Software grows in perfect atomic detail. Physically and functionally, in any direction needed, λ- calculus fills the vacuum of blind trust and digital insecurity with trusted future-safe functions.

INHERITED FAULT-TOLERANCE

The mantra for Infallible Automation and Industrial Strength Computer Science is no undetected hardware or software errors. A λ-calculus meta-machine applied by Capability Limited Addressing detects all errors on contact. Immediate detection and resolution avoid multiple mistakes. Multiple errors beset the General-Purpose Computers, making them hard to maintain. By rapidly resolving errors, the probability of a second error is approaching zero. The instinctive reaction from Capability Limited Addressing protects every individual programmed instruction. The deep and detailed checks occur at the Church- Turing Thesis's dead centre. It detects all faults, whatever their cause, whether it is hardware or software.

Adding some form of hardware redundancy, for example, the symmetrical multiprocessing architecture of PP250 or some other multicore microprocessor alternative, provides decades of hardware and software reliability without malware or hacking. PP250 readily offered five decades, much more if minimum traffic levels are relaxed. At the same time, graceful software evolution occurs without ever demanding an urgently patched upgrade or suffering any data theft or digital damage.

Transparent cyber-security is guaranteed by the digital locks, keys, and digital boundary walls of the λ-calculus context registers. Limiting cyberspace to the approved access rights of twelve typed context registers quickly detects all forms of error and prevents any attempted breakout. Security is a byproduct of focus on validating and verifying the software's functionality. No open-ended, unaddressed conditions are caused by the flawed assumption of blind trust that underpins every General-Purpose Computer.

The transparent action a program takes to unlock any access rights is performed by a tiny micro-coded routine called *'Load- Capability'* reused by any Church Instructions. It brings into sharp focus the relationship between the two alternative logical and physical forms of computation that created the Church-Turing Thesis. Pure symbolic mathematics exists on one side, while the physical (digital) structure is framed to correspond on the other.

OPEN ENDED CAPABILITIES

The PP250 design team knew an open-ended design for capability keys was critical. Capabilities are not simply a memory management alternative. Instead, they are the digital gold of computer science, where the λ-calculus is the intellectual Babylonian *art-scientist*. As civilization develops, capability keys will define everything of value to cybersociety well beyond the 21st century. Already, software tokens are vital components of blockchains and neural networks for AI.

Furthermore, to prove the point, immutable capability tokens are the foundation technology of cryptocurrencies like Bitcoin and Euritha. Capability-based computer science and capability- based computers are compatible and anchor software capabilities to the real world in easy-to-comprehend ways. This depth of this subject is plumbed in Book 3 of the trilogy on democratic cyber society.

Thus, inform capabilities are just the runtime expression of Namespace symbols that define secure abstractions of digital objects in cyberspace. The PP250 was limited to a 25-bit word size, including a parity bit to check the binary content and the memory location. So for PP250, an inform capability key was just 25 bits, as shown in Figure 16: Access Rights in Immutable Capability Keys (PP250).

The inform limitation is inadequate for the infinite variety of software-defined tokens for any chosen digital computer. For PP250, the 24-bit inform Capability Key referenced a three-word, 72-bit slot in the Namespace Table. This slot is where the Base and Limit or the remote networked geometry of the digital object is stored.

The slot is typed to hold various forms of tokens that can be defined. For PP250, it was the networked home location of an out-form Capability Key. It could also be a literal Capability Key for small two-word data objects or as the combined size of 92 bits without creating an object space. Finally, a passive Capability Key, a literal key with up to 92 bits, is an immutable value for interworking with other networked secure capability systems. These variations are summarized below, as shown in Figure 16.

The typical case is an *In-form* Capability Key to an object in local memory. The geometry of the location defines a base and limit address with a checksum to use after these values are loaded in a context register using the Load Capability microcode. Once correctly unlocked, the object is available to the computations, as limited by access rights and the address range.

Out-form Capability Keys define a network object found in cyberspace at what is called today a URL. However, a URL is a corruptible character string used through assumptions of blind trust. Unlike this dangerous binary data, unavoidable in a General-Purpose Computer and easily misused in a clickjacking attack, in a Church-Turing Machine, network addresses held as Capability Keys are immutable and prevent both clickjacking and targeted denial of service attacks through the need-to-know rules of Capability Limited Addressing.

It allows secure interworking between remote λ-calculus Namespace versions and through industry standards yet to be defined with various cryptocurrencies, wallets, blockchains, and AI tokens. The trap is set when registering on load to allow lazy evaluation. The objects invoked can be either downloaded, cached on-demand, or the request transferred for remote assessment if the trap condition fires.

A *Passive Key* is a private token that verifies permissions through an unforgeable, fraud-resistant digital credential. It is a profound secret used between function abstractions inside a Namespace. It is a fraud and forgery-resistant secret in point-to- point relationships that can encrypt and decrypt communications to prevent '*man-in-the-middle*' attacks.

A *Literal Key* holds a binary value. No memory segment needs to be allocated because of the small binary value stored in the key or the slot of the namespace table. The access rights apply to the stored value. Thus, small objects, held as literal keys, avoid delay and wasted memory space allocation. A literal Capability Key also allows the storage manager to format the Capability Key and the slots in the Namespace Table before returning a new object request.

Capability Keys decouple digital implementation from scientific functionality and logic. Consequently, each side of the Church-Turing Thesis is perfected independently, and the caveat of the Church-Turing Thesis, *'ignoring resource limitations'*, becomes fully resolved in worst-case conditions. Resource limitations do not lead to blue screen events that disrupt the computational logic or corrupt unfinished work.

The trap mechanism delays and avoids unnecessary activity, making the access rights open-ended. Capability Keys with reduced rights to limit further sharing, prevent namespace export of data and deal with watermarks to prevent Deep Fakes or other unauthorised use or misuse of a resource, for example, by adding one-time-only usage to delegated keys or transferring ownership between Threads and limiting object release conditions.

Incidentally, the PP250 also used Capability Keys without access rights to prevent spying on communication links when an object is *'moved.'* The *'Move'* instruction performs this action when downloading an object from a backing store or importing an object over a communication link.

EVOLUTIONARY STABLE SOFTWARE

As applied by a Church-Turing Machine, the Babylonian Golden Rule makes functionality and security the same problem. Measuring the error rate of objects drives a graceful process of improvement and incremental evolution. Evolution steadily improves software's functionality, security, and survival in a hostile environment. Each component has an up-to-date MTBF as a programmed function abstraction, which is recalculated if errors occur. The weakest links are transparently listed as candidates for improvement, driving each namespace's graceful evolution and adaptation to new threats.

As errors evolve, for example, caused by Artificially Intelligent Malware improvements, each namespace responds as needed. The attacks impact different software in various ways, and a valid change to one virtual machine in a General-Purpose Computer might poison another. In a Church-Turing Machine, a namespace reacts to hostility independently driven by the Namespace owner.

When a new generation of weaponised malware is launched, the threat vectors are quickly and automatically found by error reports that update the MTBF of the impacted software modules. Furthermore, attacks are traced back to a source by the watermarks of the Capability Keys in the directional DNA hierarchy for corrective and legal action. At the same time, the law and order of the λ-calculus stabilises the recovery and restoration of the qualified grade of service. The fail-safe security of the Capability Limited Addressing prevents any data loss or digital corruption, and improvements go ahead without the panic created by successful zero-day attacks in a General-Purpose Computer.

The Church-Turing Architecture mimics the Darwinian model of stable evolutionary species in nature. These software machines suffer no intolerable breakdowns because a life cycle of rebirth and regeneration restores the qualified grade of service. The directional DNA hierarchy guarantees atomic protection through immutable Capability Keys and Capability Limited Addressing, all sustained in parallel and real-time. Improvements occur in alternative ways, as error reports show new threat vectors and attacks, first, by improving the implementation of any function. Next, new function abstractions are added, and finally, the directional DNA hierarchy is changed to support the latest services in parallel to replace old functions. Centralised failures do not exist. The death of one Thread ends a flawed transaction that can be purged and regenerated if necessary.

The engineered rule of form always matching function is enforced by the symbols of the λ-calculus and a clockwork meta- machine that prevents all software interference or human sabotage. Error detection includes inside threats from undiscovered software bugs and deliberate attempts by malware and hackers to break out or break in. The data-tight rules of need- to-know and least-authority detect weaponised malware and attempted hacks on the first contact. Equally important, the recovery process dynamically reconfigures the reaction based on the current stability of the recovery experience in combination with the measured rate of the origin of errors and the owner of the namespace.

For example, the Capability Keys on any error guides the recovery process to isolate any unreliable components. Broken hardware, outdated software, new or unreliable function abstractions, and evolved attack vectors are isolated. The unreliable functions are removed by marking a slot in the Namespace Table out of service.

In background mode, the garbage collector service finds out about redundant Capability Keys. Its service finds and recycles the table slots and memory segments to clean up the aborted threads, other released items, and related resources after the recovery actions. The details are discussed later. See Figure 30.

THE PARADIGM SHIFT

By reversing von Neumann's flawed assumption of blind trust, the master-slave paradigm of General-Purpose Computer Science is turned inside-out. The atomic and cellular nature of chained function abstractions with measured reliability become faithful servants of a Namespace, an owner and a cyber society. The rapid feedback on each Capability Key keeps computations on track between the guard rails of the context-sensitive dynamic path driven by events.

Church-Turing Machines serve users instead of enslaving them to a branded dictatorship. Its profound change serves society as an elegant evolution of democratic computer science to support a virtualised society as a government by and for the people. The laws of λ-calculus regulate computation through delegated Capability Keys as event variables and the directional DNA hierarchy of a mathematical structure.

The removal of unelected superuser privileges eradicates the worst-case threats of dictatorship. The paradigm shift is from slavery under Orwellian dictators to serving citizens and life- supporting applications under the control of society. The 21st century cannot allow centralised monopolies to rule life in cyberspace. A mathematical function abstraction is a servant to a Thread, and the Threads serve the citizens of the electronic village, event by event. The qualified grade of service matches the need of the application, and there are no malware bypasses or hacker workaround that use unelected superuser powers.

Once qualified with an acceptable grade of service, the recovery system automatically sustains the approved level of service without any human involvement or best practices. The engineered design is a fail-safe public utility. Industrial Strength is the ability to survive in a hostile global environment as a dynamic digital system. It goes beyond immediate error detection to include speedy self-healing. The system reestablishes the qualified grade of service without any harm to society.

The Abacus and Slide Rule faithfully served a public need for centuries. Digital computers must do the same, without bias, fear, preference, or particular best practices and constant monthly upgrades. It is a wasteful and unreliable way to build a cyber society that only leads to digital slavery. For Ada's dream of endlessly pure mathematical logic and the science of Babbage's *Infallible Automation*, the citizens must come first. It means that data must be private, forcing migration to Church- Turing Machines. It is the only safe harbour for the 21st-century digital convergence of global electronic society.

A SCIENTIFIC SERVANT

A Church-Turing Machine is a capability-limited servant and a future-safe, ever-improving service utility—a software machine under user control. Access rights are democratically distributed and delegated, and privileges are built into the computer without unfair default. When a Capability Key is passed, some precise authority is delegated, bestowed by the innate power of one function abstraction. These Capability Keys define mathematical power but are subject to the human control of a namespace.

Said another way, the laws of society now control computer science through discrete powers and authorities of delegated Capability Keys. Obedient servants then perform delegated tasks within a user's private namespace. The delegation of power creates a new paradigm for a future-safe, democratically run cyber society. A cyber society freed from branded dictators and unelected superusers. A cyber society is run by the citizens where confidential data stays private and centralised authority is decomposed and distributed by the laws of society.

Each Thread detects and addresses threats locally through the delegated variables in a private transaction. For example, an incorrectly calculated dynamic value is detected at the point of stress. Stress must be detected on the spot, whatever the cost. Such is the case with the plane's failure of automatic flight control systems. An unforced crash can be avoided when the pilot is entirely in charge. A catastrophe occurs if errors are not detected in time or a dictatorial system will not relinquish control.

The dictatorial nature of General-Purpose Computer Science is antithetical to democracy and human reason, but Capability Keys turn computer software abstractions into faithful servants instead of capricious masters.

As a servant instead of a master, a Church-Turing Machine cannot wrestle control away from a pilot who knows a catastrophe is happening. The master-servant contradictions exemplified by the 737 MAX catastrophe will overcome 21st- century virtualised society. It is suicide to use dictatorial General- Purpose Computers that lack the means to detect software errors because the new world order places software as the most critical asset and aspect of a virtualised society, is suicide.

The paradigm shifts from General-Purpose Computer Science to a Church-Turing Machine, and the distributed Capability Keys place control in the hands of the user. Thus, delay and confusion fatal to life-supporting software applications are replaced by clear lines of capability implemented authority. The directional DNA hierarchy defines the lines of authority. A user's authority is functional, using Capability Keys that touch the surface of cyberspace in every case. The DNA strings control user functionality smoothly and seamlessly, as scientific issues with qualified, guaranteed service grades. The users control the services without the risk of a surprise or a sudden, widespread, catastrophic breakdown.

Ada programmed in this way, using clear statements of mathematical logic. A set of mechanical abstractions interpreted her statements. Her mathematical expressions defined an abstract machine. Her program, written 200 years ago, could run equally well today on a Church-Turing Machine using a mathematical namespace. Its brutal demonstration of future- safe programming is unique to meeting Babbage's standard of *Infallible Automation*. A machine of pure mathematics, unadulterated by proprietary conventions and unfair privileges. A digital implementation of Industrial Strength Computer Science that has monumental significance to the progress of civilisation beyond the 21st century.

ADA'S ENDLESS VISION

For Ada's vision of cyberspace to be realised and solve all the questions of future civilisations, digital convergence must be a success. For the electronic village to prosper and for democratic civilisations to flourish, the stable evolution of software as a servant of humanity must be made real. When the Babylonians built the physical form of the arithmetic as an Abacus, civilisation took off. World trade began, and cultures interacted peacefully. For cyber society to take off, trusted software must be a commercial reality.

The baggage of General-Purpose Computer Science prevents this peaceful acceleration into the future. At the same time, a Church-Turing Machine enables the limitless growth of separate cultures that evolve in safety alongside one another from the third world to industrial society. Computational errors, communication failures, malware attacks, mischievous hacking, and corrupt users evaporate by turning von Neumann's assumption on its head. Precisely secure algorithms running on highspeed rails as fail-safe software machines reinforce Ada's vision of the future. These software machines serve society fully and forever to save us from ourselves. Babbage was right: *Infallible Automation* is the only road to progress.

If not, society stays enslaved by General-Purpose Computer Science to opaque, monopolistic practices demanded by fickle, unruly, unreliable masters who rule digital society—corrupted by malware, imprisoned by unelected dictatorial tyrants organised by Artificially Intelligent Weaponised Malware seeking some inhuman objective.

When the general-purpose computer was pioneered, software was homegrown. The users trusted it because the user was also the programmer, but these machines were outdated due to software progress and expanding computer networks. Trust evaporates as malware, hacking, cybercrime, and corruption grow. The General-purpose computer and Computer Scientists led the nation to an unwinnable international cyberwar. It leads to a far worse future in a digitally converged electronic village where democracy and civilisation stand at the virtualization stake. Ada Lovelace envisioned the universal and extensible power of mathematical machines needed to solve all questions of science and society through Babbage's *Infallible Automation*. For equivalent reasons, Alonzo Church developed λ calculus as the universal model of computation. Despite all this science, humanity is at war with itself because of the flaws and weaknesses of General-Purpose Computers and General-Purpose Computer Science.

Alonzo Church matched the vision of Ada Lovelace using the laws of mathematical binding to scale Turing's α-machine. The combination is enshrined as the Church-Turing Thesis. As an essential relationship between mathematical functions and digital form, a monumental relationship was first discovered in Babylon. The laws of λ-calculus formalise the Babylonian Golden Rules of computer science. In every case, the most straightforward examples work best and thereby survive.

Extending into digital cyberspace, the atomic abstractions exemplified by the Abacus and the Slide Rule define a life cycle for software survival. The same survival model inherited in Mother Nature as computational genes and nuclear DNA.

This natural model processes events through life cycles, regeneration, replication, cellular threads, and variable substitution. Nuclear replication of function abstractions with variable substitution is the uniform security and scaling mechanism for life, driven by science. It works for a beehive, Rose petals, the Abacus, Slide Rules, and a Church-Turing Machine. Using dynamically bound λ-calculus with the power of substituted and anonymous variables, encapsulated mathematical functions empower and secure life in cyberspace at the same time.

Von Neumann's shared memory architecture overstretched Turing's α-machine and destroyed the Babylonian Golden Rule. The disunity between functions and forms turned General- Purpose Computers into a death trap. Piles of binary data cannot survive or resist the unavoidable wear and tear of corruption. Gross, unfair centralised privileges and branded binary conventions as centralised dictatorships make endlessly escalating cybercrimes the talisman of General-Purpose Computer Science. It is the wrong path to take for the endless future of enlightened civilisation.

CHAPTER 12

The Cults of Cyber Society

"The public character of every public servantis legitimate subject of discussion, and his fitness or unfitness for office may be fairly canvassed by any person."

Charles Babbage, 1791-1871 KH, FRS Mathematician, who pioneered flawless Infallible Automation as a clockwork mechanical, computer.

In the pre-electronic age, digital computers were batch processors before the microprocessor accelerated digital convergence. They were fragile from the start, worshipped like religious icons by Zealots who enforced the rituals of branded best practices as cult followers. These delicate digital machines were necessarily cloistered in the locked inner sanctums of mysterious secret societies.

These branded digital cults began in WWII at the beginning of the end of a pre-electronic age and the fear created by the Atomic Bomb. Throughout the ensuing Cold War, the bureaucratic cults of branded computer science were glorified and sanctified by *'timesharing'* and *'proprietary operating systems'* as General-Purpose Computer Science. In truth, science was abandoned when, in 1945, von Neumann took over from the creators of the Church-Turing Thesis in 1936.

Ironically, General-Purpose Computer Science has automated and remade the world but failed to apply Babbage's *Infallible Automation* to itself. The unstoppable force of software has deconstructed industry after industry to virtualise the world in which we live. However, software in a General-Purpose Computer is opaque and monolithic. It cannot be trusted. General-purpose computers do not achieve the scientific standards of *Infallible Automation* that motivated Charles Babbage.

Babbage used pure mathematics for *'Infallible Automation'* to remove all human mischief and mistakes, both accidental and deliberate. It is the pure, lasting, monumental form of mechanical computer science proven by the eternal Abacus and Slide Rule. He proved his flawless mechanical prototype as the Difference Engine in 1849.

However, a century later, the US Army settled on batch processing at the end of WWII. Instead of following Alonzo Church and the superior science of the λ-calculus, Babbage, the Abacus, and the Slide Rule, all united by the Church-Turing Thesis, von Neumann pushed a fatally flawed, tilted, one-sided solution to carry his name forward.

The first binary computers adopted von Neumann's overstretched physical architecture of distorted mathematics. The flawed consequence became digital cults, locked behind doors, staffed by privileged experts who wrote the code themselves. These computer designers and their teams drafted rules to be learned and performed as branded rituals, shift after shift.

Only the homegrown, hand-crafted code was approved, and errors were hidden from public view. It was the price paid for General-Purpose Computer Science. Now, after virtualizing the industrial world, General-Purpose Computer Science stays the same branded cults, inadequate for the next phase of digital convergence. A democratic society is being virtualised to share not only von Neumann's unguarded memory system; it must share an electronic village enslaved by branded dictators and exposed to international digital corruption. The democratic world rebuilt by global cyberspace is exposed internationally and subjected to undetected enemy attacks.

The digital convergence of opposing ideas between conflicted nations pollutes and corrupts cyberspace for all. The corruption of ideas and actions incapacitates critical systems. Incrementally, the privileged hardware empowers privileged malware to usurp every form of virtualised software application. Step by step, the attacks threaten freedom and equality in a democratic society. The authoritarian digital dictatorships of General-Purpose Computer Science incrementally replace virtualised democracy. In the age of globally shared cyberspace, the home-grown software that enabled timesharing to survive is replaced by open- source software mindlessly downloaded from unknown corners of the cloud. Babbage would be horrified, Ada would be discouraged, Alonzo Church would be disgusted, and Alan Turing would be disappointed.

When the microprocessor blew off the locked doors of batch processing, the Church-Turing Thesis was again ignored despite the proven advantages of the PP250. Few knew, and even less cared. A commercial race began to own the market. Science, cybersecurity, and the future were beyond concern in a race to capture market share. Now, the software in a General-Purpose Computer is under constant attack.

By day and night, international enemies, criminals, and spies move through unguarded, digital doors and unmarked software windows. Malware and hackers have open access to the soft infrastructure of 21st-century civilisation. In the unprotected digital void of virtual memory, the cogs and wheels of software are disrupted by outdated, pre-electronic age, Cold War security assumptions.

The branded cults of batch processing became the dictatorial Gods of Information Technology. Literally and figuratively, these machines were worshipped by the skilled IT staff who enforced every branded practice demanded by a supplier. The machines were worshipped by zealots who sustained the mystical practices of their digital religions. Ultimately, everything is sustained through blind trust in the infallibility of General-Purpose Computer operating system software and cult practices that replace sound science and well-engineered technology.

Cult experts in the basements of spy agencies constrained the combination of blind trust and the malice of human nature. However, even they do not keep their digital secrets safe. Edward Snowdon, Bradly Manning and whoever stole the Vault 7 secrets, published by WikiLeaks, prove beyond doubt that democracy virtualised this way is unsafe and will become dictatorial to survive.

The best effort of each cult cannot prevent things from going wrong. Crafted attacks lead these outdated computers to disgorge their secrets and crash. General-purpose computer Science has no industrial strength. Things only worked well when batch processors were locked behind doors, double-checked by experts who patched their home-grown software in real-time to reset conditions and perform cold restarts. Any problem is quickly dismissed by saying, *'The computer is down.'*

ENDLESS CONVERGENCE

Digital convergence and software virtualization are irresistible. It will continue forever. However, the digital world created by General-Purpose Computer Science has no doors to lock, no silk road to follow, and the teams of zealots needed to patch things on the spot are no more. Malware, hacking, third- party and open-source software have changed the game. In a general purpose, computer software tries to keep computations safe, but it cannot defend itself. Malware and hacks cross the functional borders and unmarked digital boundaries to silently and invisibly exploit and attack. Undiscovered, zero-day threat vectors stretch right around the world.

Shared memory, blind trust, and the WWII architecture of von Neumann cannot and will not support the industrial world. Software catastrophe is the only thing guaranteed by General- Purpose Computer Science. Life-supporting software in the 21st century cannot take this risk. The risk grows as software and hardware are added to the global network.

Massive data breaches and sudden opaque failures say everything one needs to know about the risk to democracy. Branded cults, not science, define the time-sharing operating systems built on superuser privileges. When these flawed contraptions were first sold by IBM circa 1965, the ultimate cybercrime was perpetrated on industrial society.

General-purpose computer Science cannot match the goals of the Church-Turing Thesis, the λ-calculus, and the universal model of computation. The virtualization of cyber society depends on a scientific implementation for both software and hardware. Both sides of the Church-Turing Thesis are essential for a virtualised society to survive and to meet the expectations of patriots who died for a government of the people, by the people and for the people.

Babbage's standard of *Infallible Automation* must be matched. Human interference must be prevented. Machines working at the speed of light, over open networks, and using open-source software to interconnect troubled nations cannot remain unchecked. Using software monitors, cult practices, and blind trust will lead to digital dictatorship. To sustain democracy, the citizens must be in charge. Civilisation's most precious and delicate discovery demands well-engineered, flawless science that is naturally democratic. Law and order guided by civilians.

When first switched on, a General-Purpose Computer is at once infected. Disrupted by undetected malware from the World Wide Web, the General-Purpose Computer misses the standard set by Babbage two centuries ago when his Difference Engine proved *Infallible Automation*.

In the 21st century and beyond, the institutions of cybernations must still run independently, and the software must withstand attacks day and night. Instead, they are frozen overnight by a cyber-attack. The computers for the endless future of cyber society must meet Babbage's standard if software is to survive and democracy is to flourish. General-purpose computers fail in this regard, and Industrial Strength Computer Science is the only choice for trusted software and a trusted government.

Deep in the bowels of the NSA and other security agencies, massive data breaches still take place. These organizations are paid to keep secrets but cannot do so in a virtualised world, using General-Purpose Computers. Consider the impact on a fully virtualised nation when, within a generation, by circa 2050, the software is exponentially more innovative than today, when international competition is more extreme and demand for intelligence more insatiable. Will China, Russia, North Korea, Iran and others stop attacking the American way of life, or will they use the General-Purpose Computer to bring us down?

Building the future on blind trust is just too dangerous. Not replacing this antique computer architecture inherited from WWII and the pre-electronic age is suicidal.

FLAWED TECHNOLOGY

The previously hidden software flaws came to light when the Personal Computer blew the doors off the pre-electronic age. The blue screen of death and lost work hours defined the first decade of malware and hacking created by digital convergence. Networks and open-source software invalidate von Neumann's architecture. It was ever, only fit for batch processing.

The virtual machines became a gas chamber for malware, and the unfair administrator privileges left an un-patchable gap in security. These voids, gaps, and cracks are employed by criminals and enemies worldwide, turning General-Purpose Computer Science into a weapon of mass social destruction.

Since the blue screen of death, branded layers of byzantine software have been added. They may hide the Blue Screen of Death but not cure the sickness. Sabotage continues out of sight, without warning, silently without provoking an alarm or defensive reaction. As attacks continue, hours of lost work expand to lost generational opportunity. Those who struggled to build America's experiment with a written Constitutional Democracy founded on freedom, equality, and justice are now endangered.

Awareness has grown, but the reaction is focused on band-aid patches, while the real challenge is the survival of democracy into an endless future. Abe Lincoln's words at Gettysburg on preserving America's grand experiment as a citizen's society is once again imperilled. The unelected dictators of General- Purpose Computer

Science will assume control over virtualised society. The warnings began with the Blue Screen of Death and the Morris worm unleashed on the ARPANET in November 1988. In 1994, Russian hackers stole $10 million from Citibank, showing that cybercrime pays high rewards. By 1997, a teenager penetrated the U.S. Air Force Base on Guam to flag that nothing was safe. In 1999, the Melissa virus was delivered by email, quickly infecting one in five computers worldwide, showing the overwhelming threat of digital convergence.

By 2014, Sony Pictures was taken off the air for a month or more. In 2016, a Presidential Election was subverted by a clickjacking attack distributed by a targeted email sent by Russia. Despite decades of hard work patching software, General- Purpose Computer Science is still architecturally and unavoidably insecure. The backwards-compatible, outdated hardware allows ever-smarter hacks, more serious malware, and uncontrolled cyber wars. In addition, deliberate corruption, ignorance, incompetence, and sabotage must all coexist. Layers of byzantine software limit the success rate of attacks, but shared memory, blind trust, and successful *'zero-day'* attacks still occur and are only discovered long after the event.

Underestimating the critical importance of digital integrity in the 21st century is unacceptable. Just one catastrophic attack in 100 years is too risky for a democracy to withstand, and a single breakout of Artificially Intelligent Weaponised Malware is all it takes. Some experts predict that the point of singularity when software exceeds human intelligence could arrive judiciously in a decade or more within a generation. After that, the forecast is not good.

Not since SOSP'77 have experts questioned the branded architecture of General-Purpose Computer Science. Since then, the software kernel strategy has been debunked, and the proprietary flaws in General-Purpose Computers have led the nation to the brink of disaster as a digitally converged society that sleeps with a monster.

Cloud services try to cage the monster by locking users into one cult vision. However, the cults, as digital dictatorships, conflict with the need for freedom, equality, and speedy justice in a virtualised society. As attacks exploit Artificially Intelligent Malware, it gets too late to react. It takes only one catastrophic attack to freeze a virtualised society.

The scientific solution expressed by the Church-Turing Thesis, pioneered by extreme Church-Turing Machines, continues to be ignored. Charles Babbage and Alonzo Church must be turning in their graves. The full adherence to science, through Infallible Automation, thwarts the problems of cybercrime and corrupt users.

SPOOKS AND SPIES

Sadly, spooks and spies at the NSA and their foreign friends still live in the Cold War. They foster General-Purpose Computer Science to empower their weaponised malware that spies on the world. Monopoly suppliers collude to collect unfairly obtained data from private citizens.

This relationship has no interest in changing course, even though these agencies must also serve society. To serve the citizens of the 21st-century cyber society, the government must defend the nation from attack and external domination. While everything else in the world was turned upside down and inside out by the General-Purpose Computer, these machines are still stuck in the past. As pre-electronic age relics began after WWII, they accepted the Cold War, the cult mentality of identity-limited cybersecurity. The teachings of the Church-Turing Thesis are still ignored. These platforms do not serve the citizens because they serve neither science nor society.

Dramatic changes are needed for mathematical science to keep us all safe from our inhumanity. The impact of digital convergence on future society cannot be ignored. The pursuit of unbridled commercialism by unelected monopolies is a rerun of the 19th Century. It makes regulation unavoidable, and the recent problems with data privacy at Facebook, Google, Amazon, and Apple make this clear. However, these monopolies flourished because dictatorial powers for market domination were built into the foundation of General-Purpose Computer Science. Government regulations are just extra patches that will not address the scientific root cause.

The wealth of nations and the standards of civilisation will fall if the cult solutions to computer science remain. Life in the 21st century must prevent industrial dictators, malware, and hacking. Computers and software must be democratised to be trusted. A level hardware playing field is the first regulation that is needed. Dictatorial cults are unfit to run the digitally converged eternity of cyberspace. The silence of scientists who must know this to be true promulgates this failing system. If all this was out of sight in 1945, when the Church-Turing Thesis and λ-calculus were ignored, and again in 1977, when Capability Computers were dismissed, it is as clear as day. The evolution from homegrown batch processing to unsolved global cyberwar is a scientific and industrial failure. Intel, AMD, ARM, and other hardware suppliers must be regulated to implement computer science.

MALWARE D-DAY

The current cybercrime is the failure of government action. Cybercrime is the fastest-growing profession worldwide. Enticed by the market size for undetected and underreported cybercrimes with very high payoffs, crime will continue to grow. It is the most profitable occupation for underdeveloped societies. Meanwhile, cyberwar continues to steal from the industrial world, and AI continues to progress towards the doomsday of singularity when super-smart software rules society.

Our enemies, like Russia, North Korea, Iran, and China, are not playing games. They wish to destroy Western society. Furthermore, governments, spy agencies, criminal gangs, third- world slum dwellers, teenage hackers, as well as crude marketers all have their hands in spying. They all use weaponised malware bought, copied, or stolen from expert sources, including the treasured tool kits of homeland security now freely available from Vault 7 at WikiLeaks. The dark side of humanity is winning in Cyberspace because the software is defenceless.

Freedom is under attack, and authoritarian manipulators are in charge. The virtual machines in a General-Purpose Computer are the battlefields. Digital landmines hide everywhere for every conceivable criminal, business, or military purpose. When this malware detonates, the General-Purpose Computer ignores the explosion. How unscientific is that?

Software errors must be detected by hardware. Babbage's hallmark of *Infallible Automation* must be achieved, not ignored. The changeover starts in hardware. Instead of serving society, these cult machines only serve the skilled and powerful engaged in unseen, undetected manipulation, crime, and corruption. The undetected attacks occur in a soundless digital void, where criminals hide their tracks while stealing the nation's treasure. The stolen information vanishes to a remote corner of the electronic world.

Civil law and order, the cornerstones of democracy, are missing in General-Purpose Computer Science. Trust evaporates in this unsavoury environment, and criminals become bolder as their crimes are unpunished. Damage is unnoticed for months or years until detected externally, sometimes by the press or WikiLeaks publications, others by the impacted citizens, but nothing constructive occurs.

Meanwhile, enemy states are preparing for their Malware D- day. When ready, the landing grounds are soft, and hidden forces are in place. They cross the unmarked boundaries to occupy the high ground in advance. When the attacks start, it will end in a flash without any way to fight back. Artificially Intelligent Weaponised Malware will be at work, the police will fail to function, as will the armed forces, drones will not fly, banks will fail, supply chains will break, shelves will be empty, and the government will be helpless. The preservation of a virtualised nation depends on computer science to sustain trust. It is time for society as a democracy to regulate industrial strength in computer science.

DIGITAL SARIN GAS

Weaponised malware is amplified by artificial intelligence. It pervades global cyberspace as a colourless, odourless, undetected nerve agent. The purveyors of these deliberate crimes against humanity in the electronic village should be prosecuted to the maximum. It has extreme potency. It is the nerve agent of the 21st century and beyond. The gas chambers are the overstretched virtual machines from WWII used by von Neumann for batch processing. Its outdated hardware architecture is dysfunctional regarding software, and as a result, civilised industrial progress is stunted.

False News and Deep Fakes are byproducts of undetected crimes and lawless software in cyberspace. Undetected crimes destroy all trust in computers as a platform for civilised progress. Prosperity falls as existential threats grow, impacting everyone and everything, from digital supply chains that crisscross the electronic village to citizenship and democracy, undercut by cult monopolies.

This lack of integrity destabilises national security, hurts and harms all citizens and undermines the government. As the weaponised, artificially intelligent malware discombobulates the virtual machines of General-Purpose Computer Science, grief and anger spill out of cyberspace to discombobulate society.

A never-ending stream of backwards-compatible patched upgrades further disrupts forward progress. The digital world is facing the wrong direction. Every patched update comes with incoherent, conflicting new cult practices and more autocratic, enclosed clouds.

Suppliers have decided to go it alone in their best interests and capture clients in proprietary prisons to be spied on, selling their never-ending personal histories to feed their corporate success. It is the guise for cybersecurity in the 21st century.

Meanwhile, criminals and enemies are hard at work refining their killer D-day landings for malware to sweep in unnoticed and silently fill the cyber voids with digital sarin gas. Every virtual machine is at risk, and while the unelected cult dictators wield absolute power, they still cannot detect or prevent the next zero- day attack. Paradoxically, the absolute power granted by the default rings of privilege is now outsourced to foreign administrators in some of the most corrupt and dangerous places on earth. No one considers how bad it will get.

CLICK JACK ATTACKS

A General-Purpose Computer is overwhelmed by a simple email attack. Estimates show innocent mouse clicks activate 90% of all attacks. While civilians drive cars protected by seat belts and safety bags, the equally dangerous General-Purpose Computer offers no such civilian protection beyond leaking firewalls and incompetent virus scanners.

There is no patched solution to assumptions of blind trust and superuser privileges. Malware and hackers succeed by freely crossing boundaries and creating fraudulent binary data. Some experts make the excuse that social engineering cannot be fixed by computer science. The argument does not appreciate how crucial civilian trust is to the civil progress of society.

Telecommunications automation was born from a lack of trust in the manual switchboard operator[94]—another example of unfair superuser privileges. Every network connection, email, instant message, and whenever a user touches a keyboard, there are threats that criminals exploit. All these threats are resolved by science and soundly engineered technology.

---

94   Strowger's was an undertaker. His business started losing clients when a competitor wife used her unfair privileges as an operator at the telephone switchboard to direct Strowger's clients to her husband. Motivated to level the playing field he invented the automatic telephone exchange, receiving a patent in 1891. It is said he constructed a prototype of his rotary switch using his hat box and hair pins provided by his wife.

Cybersecurity must include every keystroke, every mouse click, and every other event to be caught *red-handed* as cybercrimes start. Humanity is at war, fighting with itself, and computer science can proactively help to solve the problem through Babbage's standard of *Infallible Automation*.

A virulent breed of ransomware attacks froze the healthcare services in the United Kingdom and other services throughout Europe and, more recently, in the USA, culminating in the attack on New Orleans that prompted a State of Emergency to be declared in December 2019[95]. The software cannot be nursed day and night by experts who created these concoctions and then demand they be turned off to prevent a catastrophe. A corrupt user like Edward Snowden or Bradley Manning can still walk out the door carrying the nation's private secrets, all loaded on a thumb drive. Meanwhile, spy programs lurk in every machine, eavesdropping on cameras, microphones, and keystrokes and talking to a distant home base.

With the Internet of Things, the list will grow. Every door and window, car piston and brake pedal become objects to attack. This process of digital convergence accelerates the total virtualization and observation of life from birth to death, including every institution of good government. Everything cannot be exposed this way to catastrophic failure. The electronically governed society of the 21st century and beyond is at an unacceptable risk of failure. The nation must improve over time and not die in the process.

The prime targets must be hardened; the supply grids for electricity, food and water, government systems, and law and order are top priorities. Trusted software for life-supporting services and an endless need for Industrial Strength Computer Science must be commercialised.

Weaponised malware is deadly, efficient, fast, and hard to trace; it is the weapon of choice in all future conflicts. The war will end one day in the future in an instant if things do not change. Nothing

---

95   Mayor LaToya Cantrell told WWL-TV that she expected the cost of the attack would exceed the $3 million cyber insurance policy the city has in place, and that she will seek to increase the policy to $10 million next year. She did not discuss the insurance policy at the briefing.

will work when nations immersed in the electronic village awake, paralysed, with computers frozen, by international ransomware delivered by one of our enemies[96]. Dry runs to this effect occurred in May 2017 when ransomware encrypted Microsoft Operating Systems across Western Europe[97].

It is an impossible situation to accept. Even after an upgrade, new threats appear, and the upgrade must be installed on every computer, one by one, worldwide. It is complex, costly, unpleasant, incomplete, and always too late, and such practices are ineffective on any comprehensive scale due to a lack of skilled staff. Furthermore, new practices must be applied during installation. It is non-trivial since it concerns the earlier unique history. Its worldwide condition is uncertain, increasing the burden of painful and debilitating practices that only serve to pass the blame from the suppliers to innocent civilians.

Supplier best practices are scapegoats for the lack of Industrial Strength in General-Purpose Computer Science and the suppliers' unwillingness or inability to resolve these threats. By blaming the user, they duck the responsibility and keep their unfair grip on the market. It must be noted that the pilot is the first one blamed in any crash, no matter how incompetent the computer-automated flight controls might be.

Finally, the unelected superusers cannot be ignored in the threats to virtualised democracy. It is a fatal flaw in a cyber society. The superuser is at odds with the separation of power in society. As ever more services are virtualised, the dictatorial power of unelected system administrators is the ultimate, unacceptable threat to freedom, equality, and justice. Its absolute, unguarded power can destroy a nation, piloting democracy vertically into dictatorship like the unfortunate 737 MAX.

Early warning bells are ringing loudly. Corrupt administrators, any privileged user at the right keyboard, or any malware in the right place can spy on and disrupt the electronic village. As these risks grow,

---

96   FBI director warns that Chinese hackers are preparing to 'wreak havoc' on US critical infrastructure

97   The growing threat of ransomware - Microsoft On the Issues

the tasks are outsourced overseas. The future is grim when corrupt locations offer the lowest cost as a strategy to spy on America. They gain the knowledge and power to destroy everything of value to society. It includes the survival of democracy. The virtualization of a free society cannot be so tragically misled.

A patch can fix a known flaw in the software, but it cannot correct the unfair hardware privileges of General-Purpose Computer Science. Furthermore, long after the critical event, patching cannot recover the treasure already lost before the attack was known. The software architecture in a General- Purpose Computer is full of threats from a wide range of unfair privileges created by a blind trust, and the patched upgrades are always too late and incomplete. The biggest threats are always found 'on the job' and have never previously been addressed.

Innocent citizens expect computer scientists to do their job of faithfully applying the laws of science and nature to engineer reliable, fail-safe, up-to-date technology. However, outdated, incompetent, backwards-compatible General-Purpose Computers are all they offer.

OPAQUE CORRUPTION

The progress of Civilisation cannot survive this opaque nonsense; industrial nations cannot risk the consequences. Society has already forgotten how things worked before the PC, but no one knows why things fail in General-Purpose Computer networks. Finding expert help and permanent solutions is hopeless, and advice supplied frequently makes things worse. Under close questioning, even security experts admit they are selling snake oil.

Unfair privileges are the root cause of every successful attack. It is the same with a clickjack or a data breach. The users do not cause this; the 'best practice' rule - *do not click* on unknown links or open strange emails is in direct opposition to the reason for digital convergence. Research, education, and learning are actions that enter the unknown, and cyberspace must defend individuals struggling to learn.

The endless electronic society demands solutions to solve problems instead of creating problems. Computers serving the long-term interests of the citizens do not exist. The user is not careless; instead, the computer scientists who promulgate the von Neumann WWII architecture are thoughtless. Blind trust in shared memory, privileged operating systems, and corrupt users define the problem computer science must solve. Only the best will satisfy the future.

The mumbo jumbo of these cult concoctions can transport us to the far side of Mars while stealing our identity. The software can replay the past in full motion as virtual reality, but without warning, it will empty one's bank account. Computers can show us the future or turn off the power grid in a flash, and programs that inspect nature in anatomic detail again using virtual reality can encrypt the hard drive equally well and demand a ransom to get back on the air. Indeed, the added overlay of augmented reality and artificial intelligence will quickly turn against society through the unseen corruption of someone's hard work. A user interface that hides a click-jacked button with false text and fraudulent images is only the start. Forged video streams from highjacked providers misrepresent the truth while hiding devious, socially engineered attacks that move beyond cyberspace into the natural world.

Nothing is guaranteed because opaque corruption starts at the deepest level of General-Purpose Computer Science. Individuals can reach across space, throughout time and beyond imagination to educate, share, socialise, and guide. In contrast, others mislead, confuse, steal, corrupt, or worse, sabotage industries, governments, elections, and national utilities. They do all the above without warning, detection, or the risk of punishment. We do not know where this is taking us, and the challenge for a democratic society and good government is to protect us from this future that only leads to Orwellian dictatorship.

CHAPTER 13

On Digital Enlightenment

'Forget this world and all its troubles and
if possible, its multitudinous Charlatans--
everything in short but
The Enchantress of Numbers.
Ada Lovelace

For decades, the two sides of the Church-Turing Thesis were disconnected as a hypothesis without proof. The caveat of the Church-Turing Thesis dismisses the incoherence by *'ignoring physical resource limitations'* see Chapter 8 - The Clockworks of λ-Calculus. However, neither nature's universe nor Ada's dream has caveats. It is the job of computer engineers and scientists to resolve these questions. The vision, now called cyberspace, is nature's mathematics as a universal model of computation. Babbage solved all his resource problems, as he proved this with *Infallible Automation*. Ada's dream is achieved by marrying the hardware and software sides of the Church-Turing Thesis to solve any disparity by enforcing the law of the λ-calculus.

The caveat exists in monolithic computer science's imperfect, pre-electronic age view. The solution is found in the atomic medium of the nuclear age that led to the discovery of the point transistor, object-oriented programming, DNA, and the microprocessor. Software functionality is dynamically bound as atoms using the Babylonian Golden Rule, uniting form with function atomically at the level of individual mathematical symbols. Uniting form with function aligns every aspect of theory and practice to make the caveat malware and superuser privileges disappear.

As in nature, functionality and form coincide with security, power, and authority. The rules of natural selection make it so. When bound together as one, the atomic simplicity pioneered for computers by the Abacus, the Slide Rule, and the Difference Engine matching boundaries unite the solution. Atomic binding is the spirit of the electronic age, and the Babylonian Golden Rule makes all the difference to effective functional solutions that survive the tests of time.

Using the λ-calculus transparently separates concerns to solve the dilemma incrementally. The pre-electronic age view of computer science is physical and monolithic, whereas the electronic age must be abstract, atomic, and functional. The General-Purpose Computer is focused on abstracting physical memory and centralised control, while instead, the Church-Turing Machine concentrates on abstracting functional science as distributed control with inherent security.

The wrong starting point in the pre-electronic age created the caveat because of the unsolvable problems with General- Purpose Computers. The Babylonian Golden Rule brings together the two halves of the Church-Turing Thesis. They are not monolithic and independent; they must work together as the yin and yang of computer science. Virtual memory took the General- Purpose Computer in the wrong direction, building isolated, independent procedures. The pre-electronic age drove a need for physical boundaries, locked rooms, batch processing, and the Cold War security model of Identity-Based Access Control. However, the electronic age solution is abstract, atomic, integrated, and united. Atomically, pure mathematics, as bound by the λ-calculus, is the materialization of software from scientific symbols. When both sides of the Church-Turing Thesis reference the same atomic substance of the electronic age, there are no orthogonal resource caveats to dismiss. When the form is united with the function, the power of pure mathematics is achieved in the real world as a fail-safe machine. A world bound atomically where each problem is elemental and solved in total.

A Church-Turing Machine mates these two views together as one. The caveat and the deconstruction of the monolithic, pre- electronic age, centralised dictatorial views of General-Purpose Computer Science are decomposed. The model updates Identity- Based Access Control with Capability Limited Addressing to solve cybersecurity problems forever.

At the start, Turing's α-machine was overstretched into the General-Purpose Computer, and λ-calculus was ignored. But the atomic machinery of a Church-Turing machine enforces the λ- calculus as the stable, dynamically bound, fail-safe and future- safe computer science. It is the solution Babbage uses for *Infallible Automation*. It mimics nature's ability to build complex living organisms from pure mathematics, as demonstrated by bees that instinctively build the pure hexagonal shapes of their mathematical honeycombs.

The resource caveat applies to General-Purpose Computer Science because it is a monolithic fabrication. In that sense, it is unreal. Not even volcanic rocks are monolithic. Nature is built atomically; even atoms are fabricated from sub-atomic and mathematical particles. A Church-Turing Machine solves the caveat because it is atomic, and the two disconnected views on hardware and software that beset the pre-electronic age disappear.

When mathematical theory matches the implemented solution, everything remains pure. In this perfect scientific solution, the laws of computer science are obeyed. Bad actors only win when powers can be misused because the playing field is tilted. When cybersecurity is guaranteed, and the citizens of society assume control of cyberspace, the future is managed transparently through the universally fair, level playing field of mathematical science, logical form and encapsulated functionality.

ATOMIC RESPONSIBILITIES

As Alonzo Church conceived a λ-calculus function abstraction, it is a symbolic mathematical construction, an expression of pure science. As a reusable mathematical function, when characterised by some variable instance data, it becomes a λ- calculus Application, a frame in a Thread executing the universal model of computation. The tasks are a chain of functions, and the data are orthogonal atomic components forming a live thread that is born and eventually dies.

For example, data and functions can exist together but in infinitely different ways. In the thread of life for a beehive, some bees collect honey, others build the hive, and in one particular case, a single Queen lays eggs. Yet they all share the DNA of the specific bee hive.

The function abstraction of a bee is a reproducible bee machine, but reproduction is a reserved function. All females might be capable, but only one has the right to produce. Each egg has a DNA string. It defines the bee machine in total, but when combined with the instance data, the machine of life is characterised, framed function by function as an individual bee by the universal model of computation. Each bee is specialised to work every aspect of life needed for the hive to thrive.

Individual bees have a life cycle, but a lost bee does not threaten the hive's survival. Worker bees only last for five to seven weeks, from the first few days working inside the colony to the last ones for gathering food, nectar, and pollen, while the queen bee, free from disease, can reproduce 1,500 eggs a day for three to five years.

Yet the life of the colony continues when a queen bee is replaced. A queen releases a smell to stop other female ovaries from functioning. When she dies, the signals stop allowing other female bees to lay eggs. A few larvae in *'queen cells'* are fed royal jelly, and one ascends to the egg-laying throne after killing all competition. During this month-long replacement cycle, the hive is chaotic and vulnerable, but the colony still survives.

The bee instances sharing the same DNA are the same generation of bee machines. They work together to build the future using their private instance data within the machine. The functions defined are democratically shared through the directional DNA hierarchy. The DNA machine is perfectly engineered for bees to live together as a colony that survives, queen after queen, generation after generation, as an independent bee society without a centralised super privileged, royal dictatorship or any centralised authoritarian government.

Once again, there is no central operating system in nature's universal model of computation. The queen bee does not run the hive. The worker bees, the bee citizens, control the queen. While the queen is pivotal, she is not in charge. The laws of the λ- calculus and the laws that operate life in a beehive and life in the living Threads of a Church-Turing Machine are all the same. The λ-calculus Variables, Abstractions, Functions, and Applications of the beehive are dynamically bound atomically and anatomically from pure mathematics that drives everything in nature.

In computer science, it is the λ-calculus and Capability Limited Addressing that maps mathematics from the purely logical world into an atomic physical world, where every problem, including the death of a queen, has an anatomically defined, scientific solution that can be reduced to mathematics on a university chalkboard. In computer science, the λ-calculus is nature's universal model of computation as a programmable mathematical machine—a machine where data and functions are separate and distinct.

The Namespace is a beehive for the DNA of a machine defined by a species of software. In a Namespace, the Threads of computation are the individual bees working from a common DNA specialised by their own privately substituted variables. The Threads use function abstractions chosen individually from the directional DNA hierarchy. Its individuality is the choice of the citizens, not a royal dictator.

SOCIAL RESPONSIBILITIES

In a human world driven by the λ-calculus, service providers are unlike Facebook or Google. Instead, they are limited in providing function abstractions. They do not control or own the confidential data that always belongs to the citizen Threads, secured in a Namespace. Individuals must be elected to serve in central functions like a queen bee. It is the essential mechanism of democracy. These central functions are the abstractions of cyber society. Candidates are chosen for their refined ability, but they do not rule as dictators or have open access to confidential data.

The decomposition of the dictatorial General-Purpose Computer architecture changes everything, not just eliminating malware and hacking. Instead, the citizens of a cyber colony keep data in a secure Namespace with an independent mind and a different life cycle. A colony is a social group defined by a community's identities and confidential data as the citizens of another Namespace. Each Namespace as a colony has capability tokens that define function abstractions as DNA strings. The citizens remain in total control of their data.

As Namespace owners, citizens decide what software functions to install and what data to share. Still, confidential data in a namespace cannot be accessed without the permission of an owning citizen. The namespace owners of a social group are elected to control access to the democratic function of a cybersociety by the citizens for a limited time and in a restricted way.

A λ-calculus Namespace is like a beehive where the Threads are the bees sharing a social responsibility to the hive as a colony. Its universal model of computation is the paradigm for 21st- century cyberspace. The General-Purpose Computer is a privileged digital dictator that tilts the playing field to aid and abet both crime and corruption. These computers are no safer today than at SOSP'77 when Butler Lampson claimed incredible but unrealistic software advantages of centralized, superuser operating systems that will never be realised. He forestalled the progress of computer science by another half a century.

The universal model of computation reverses the mistakes of General-Purpose Computer Science. Mathematics as symbols are functional servants of science and society. The policy of the DNA reaches down to the computational surface of computer science through the immutable Capability Keys that define a λ-calculus Namespace. A directional DNA hierarchy establishes a policy for life in the Namespace, and each Namespace has a private policy as another species of software. A functional species of software that controls digital security as a policy. The guarded, graded access rights extrapolate from atomic interactions into functional, socially responsible interactions for each colony.

These functional extensions keep citizens in control. It is essential for the good of the 21st century. The people must still govern data privacy and constitutional democracy. The citizens who live and survive in a cyber society will extend their preferences and personalities into cyberspace through their private namespace as an attached digital shadow using the individual DNA of the cyber citizen.

War is an unfortunate side of humanity; it will never change. However, the General-Purpose Computer encourages this dark side by expanding the battleground of attack worldwide. Because these attacks are undetected, they are also unresolved and move beyond cyberspace into the natural world. When attached equipment is destroyed, the tribal instincts of humanity place blame on outsiders as foreigners and trust in national governments fades. Threats are met with counter-threats, suspicion becomes fact, and false news fills the void of the guesswork that replaced science.

Attacks by unknowns, acting alone or in teams, are located anywhere in the world. As crimes lead to conflict, Ada's dream of a world empowered by mathematics is destroyed. The world is falling apart from the lack of science in the General-Purpose Computer. General-purpose computer science is an oxymoron, misdirecting the future of civilization.

Democracy is transparent. Equality is founded on freedom and justice under the laws of the land. The citizens decide the rules, as in society, as found in a beehive. Instead, a General- Purpose Computer is tilted in favour of suppliers and a criminal class led by spy agencies who prevent serious competitive alternatives from taking off. Identity-limited passwords are the only protection in a General-Purpose Computer, but passwords to virtual machines solve the wrong problem.

The software backdoors and user downloads remain unchallenged. The virtual machines became malware gas chambers that could not protect software or detect hacking and malware. The General-Purpose Computer does not understand the λ-calculus of life. The individuality that builds the social fabric of a beehive as individuals with cellular boundaries and the private individuality of a citizen are washed away by compilers. The monolithic result is inefficient and exposed. Digital infections misuse unfair privileges that cannot be fixed by patching.

MONUMENTAL COMPUTER SCIENCE

Monumental computers, like the Abacus and the Slide Rule, last forever by abstracting mathematics as an inner machine. Every lifeform, including the bee, is a monumental, survivable form of computer science. Since mathematics is future-proof, so are these machines. They have mathematical integrity and a generational life cycle. As machines, they are defined by the blueprint of a reproducible DNA. As such, they survive as abstractions of mathematics, not abstractions of hardware as with virtual memory.

The distinction is critical. Symbolic computations are defined as living species, not as static machines. Symbolic expressions are taught to children from a chalkboard at school. It is a level playing field, expressed peer-to-peer, symbol-to-symbol. The students are the computers, programmed by the teacher, writing on the chalkboard. Von Neumann's monolithic virtual machines are dead to life. It cannot understand the mathematics on a chalkboard. It does not comprehend mathematical abstractions that separate implementations from theory to survive and scale.

The life cycle of a worker bee, a queen bee, and a beehive are different, but the machinery's core cellular chemistry is the same. The real difference is hidden in the DNA structure. Making the DNA transparent is vital to the survival of cyber society in a virtualised Democracy. The universal model of computation drives every living thing with a monumental form of computer science built atomically through chained cellular structures with rigid digital boundaries limiting computations to simple mathematical statements as found on a virtual chalkboard and framed as symbolic statements that can not only be executed by digital mechanics but also be thoroughly tested, function by function.

An endless, natural science of individuality replaces blind trust. A Church-Turing Machine detects and prevents harm. The technology stack is founded on a λ-calculus meta-machine, enforced by Capability Limited Addressing, that encapsulates Turing's α-machine as a fail-safe computational framework for a species that lives in a private namespace.

Cybersociety can trust the software that runs in a Namespace to last generation after generation, as a colony of threads, like the bees in a beehive. The individuality of the beehive is defined by the colony of bees that individually perform roles that allow the colony to survive. Furthermore, individuality is always private because the computational threads (the bees) belong to a colony of confidential data and functional DNA. The data and the functions as abstractions all observe the Babylonian Golden Rule of atomic form and function.

THE VIRTUAL CHALKBOARD

I t is not the function so much as the structure empowers survival. The survival of a democratic society depends on the universal model of computation, just as the survival of a colony of bees uses the universal model of computation to survive. The criminal interests that penetrate General-Purpose Computer see theft as their primary interest, but enemies go far further. They attack the democratic way of life. Anything unfair in computer science is used to that end.

A Church-Turing Machine is a student in a classroom executing mathematical expressions. The student will not live forever, but the education system will, like mathematics, pass the tests-of-time. When Ada Lovelace wrote the first software program circa 1843, she used the same syntax still used in classrooms today. Her program has passed the test of time, and if implemented functionally instead of physically, it still works today.

Her symbolic expressions do not age. Her future-safe program deserves a future-safe computer. The unfair privileges threatening virtualised society vanish when power is functionally and individually distributed in a Namespace. Power is firmly transferred to individual citizens who colonise cyberspace. Colonization is a social characteristic vital for cyber-democracy to live and survive in a hostile global network. Networked computer software must act for society and not independently from society. Cultures demand private cyber colonies structured in their way, with DNA for their form of cyber society. They can all exist at once, each in their namespace. All play on a level field of pure functional science that removes all dictatorships and detects every criminal.

On one side of a schoolroom, the teacher's side of the chalkboard, there is pure functionality, the symbols, and the functions. On the other side of the schoolroom are the students as human computers. The students, as computers, must understand each and every symbol. Each symbol has a meaning interpreted to obtain a result in the child's mind. The children embody the Church-Turing Thesis in action as Church-Turing Machines. The digital media in a digital computer can do the same as a child. It is the shortest line between the concept and the realization, a straight line that leaves all the baggage of the General-Purpose Computer behind.

The symbols are, at once, tokens of authority and functional power. In a beehive, the queen's power is a smell that goes hand in hand with her egg production. In a cyberdemocracy, power must be recognised by the citizens, and the citizens must elect the powerful. However, as in a beehive, the powerful are not in charge of anything more than a narrow and specialised functional task for a limited time.

The tangible, immutable strength of a Capability Key extends the hands of a user, as citizens of society, through directional Namespace (as DNA hierarchies) down to the computational surface of a Church-Turing Machine where all actions occur.

These Capability Keys are the eternal servants of humanity; they rule access to cyberspace on behalf of the namespace owner. Criminals and dictators are powerless and isolated by default. A trusted Church-Turing Machine is a commercial solution for trusted software. It is a scientific solution that survives like the Abacus universally and endlessly, generation after generation, as a faithful, flawless servant.

The architecture delivers Babbage's flawless standard with Ada's endless vision and none of von Neumann's or Lampson's downsides. Hardened digital security protects software while eliminating digital corruption even in unattended, hostile situations. It is a monumental solution for computer science, vital to the future progress of interconnected but divergent cyber societies. In the age of global digital convergence, society's forward momentum depends on the perfection of functional cyberspace.

Babbage understood the harm to society from human errors and showed the way with the flawless mechanical engines he wished '*to be driven by steam*,' not by humans. At the same time, Ada Lovelace envisioned his *Infallible Automation* solving every question of humanity in the natural universe. When Alonzo Church established his theory that underpins Babbage's standard and Ada's vision, his talented student Alan Turing explained the gene for his universal model of computation. By shining a mathematical light on the human errors that frustrated Babbage, the deliberate errors intended to cripple cybersociety are also resolved.

Thus, a Church-Turing Machine is a monumental solution like the Abacus. It is the ultimate form of computer science that, like the Slide Rule, has worked for all mathematics over time using the same Golden Rules that fit the ages, spanning the world to the last millennia.

INFALLIBLE AUTOMATION

F lawless computer science was Babbage's dream; he was frustrated by the slightest error in the least significant digit of the tables produced by his students. His perfection inspired Ada's vision as the universal means to understand everything in life, imagination, and nature. Their ideas led to Alonzo's discovery of λ-calculus as the universal model of computation and Turing's α- machine as a digital gene for computing a function abstraction, a digital version of Richard Dawkins' Selfish Gene[98], a genetic survival machine.

A Church-Turing Machine is the best of breed, a monumentally significant digital computer, a Dream Machine for the 21st century and beyond. The eternal solution to *Infallible Automation*. The universal language of numbers and functions chain together as pure, hardened expressions computed symbolically, precisely as taught on a chalkboard at a school or university. The symbols form as hardened object-oriented machine codes are assembled and reassembled dynamically on the fly. One by one, these atomic functions build the cells, organs, limbs, senses, body, and mind of a mathematically programmed software species.

A software machine is directly connected to the real world through an infinite variety of digital transducers connected internationally, as in the telephone age and as it is again for the Internet of Things. Like animals surviving a hostile wild world, the digital environment of

---

98   The Selfish Gene is a 1976 book by Richard Dawkins on evolution, that builds upon the principal theory of George C. Williams's Adaptation and Natural Selection (1966). Dawkins term "selfish gene" expresses a gene-centred view of evolution, this view argues that the more individuals are related, the more willing they are to cooperate. See also The Selfish Gene – Wikipedia.

wild software, like animals, needs the instinct to survive, reproduce, and evolve. In a Church-Turing Machine, the survival machinery is crosschecked by Alan Turing's α-machine harnessed to and physically encapsulated by Alonzo Church's λ-calculus meta-machine.

Malware and hackers have no default privileges to misuse. When downloaded, malware is dynamically confined to the digital boundaries of the download. As such, it cannot cause harm to other software. Doing more requires Capability Keys beyond the ones created by the download, and any attempt to break down is detected and prevented. These extra Capability Keys must be negotiated, and any misuse is immediately detected. The framework remains stable by detecting and isolating errors while accommodating new functionality with a revised DNA as an improved machine.

A Namespace is a digital extension of individuals and society with a clearly expressed purpose, engineered boundaries, and survival ability. It is a pure, unadulterated form of computer science. It uses the universal model of computation that is scientifically trustworthy, achieving Babbage's standard of *Infallible Automation* with Ada's vision of everlasting enveloping computer science that peacefully solves every question of humanity forever.

SYMBOLIC ADDRESSING

Unlike Turing's statically bound imperative procedures, a Church-Turing Machine is functional. Six Church Instructions add symbolic binding to build function abstraction from Turing-Commands. The Turing-Commands are bound to symbols in the expressions written on a virtual chalkboard. They are not bound to pages in static linear memory. The symbols on the chalkboard are expressions like those taught in school and university. No default power exists to overrule the expressions on the chalkboard.

The twenty-five steps Ada wrote to solve the Bernoulli Numbers reapply. Consider the accelerated progress of civilisation when programs like Ada's all last forever. It is the value of industrial strength in computer science that survives. The Enchantress of Numbers, Ada Lovelace, as Babbage called her, understood the faithful calculation of a function. A square or a cube root function serves humanity only if they faithfully pass every test of time.

By framing mathematical functions as protected atomic computation, the algorithm and the sub-atomic media are bound to the frame of a computational Thread. The actual storage locations are unimportant. Capability Limited Addressing offers this flexibility, along with cybersecurity. Church Instructions manage the transition between frames guided by an *EnterOnly* Capability Key. The *EnterOnly* Capability Key unlocks a transparent door to the digital geometry of a private Minotaur's labyrinth. The labyrinth materialises the bounded functionality of one link in infinitely strong chains of functional mathematics.

Dynamic binding is a formal mechanism for symbolic computation implemented by Capability Limited Addressing. The λ-calculus meta-machine mechanically frames the computational surface of cyberspace. On one side is pure mathematics, expressed symbolically on a virtual chalkboard without distortion. On the other side is the atomic digital media materializing each symbol as a function.

Symbolic addressing obviates the need for compiled virtual memory and all the attendant cost and complexity of dangerous linear page registers, rings of dictatorial privilege, and central operating systems with single points of failure. Instead, memory is indirectly translated by Capability Keys and Capability Limited Addressing. It is a secure alternative to information accessed on the World Wide Web. As a single universal, cyber-secure addressing mechanism, it removes all reason for a page-based virtual memory and all the threats created by the internet.

To emphasise this, consider a Church-Turing Machine creating a page-based abstraction that simulates virtual memory. Indeed, anything else in a General-Purpose Computer. The reverse is not possible. It not only smooths the migration strategy of various cult solutions but also demonstrates the superior advantages of the Church-Turing Machine. The reverse is impossible because a General-Purpose Computer is statically compiled. The statically bound imperative Turing-Commands cannot operate symbolically.

SOFTWARE MODULARITY

Switching from the shared linear storage of WWII to symbolic addressing is easy. The program-controlled context registers replace the opaque page registers and displace the operating system. Just six Church Instructions remove all the hazardous baggage of General-Purpose Computer Science. The change to the instruction format trivially redefines the page address as a λ- calculus machine context register from Figure 12. Under program control, the context registers cache the object type, access rights, address boundaries, and status of the current computation frame. The hidden page registers disappear. The most straightforward alternative instruction formats are shown in Figure 26.

General-Purpose Computer	Compiled Image	Shared Applications	Shared Programs	Binary Register	Binary data
Church-Turing Machine	Namespace	Abstraction	Function	Capability Context	Information

FIGURE 26: THE INSTRUCTIONS FORMAT CHANGES BETWEEN COMPUTERS

Given a longer word length, the functionality and security can be improved by defining the needed Capability Keys (by context register ID) in a Church Instruction request. The context registers then scope the computation as both media-tight and data-tight. When the scope is precise, tailored to the context of accessible symbols, one function abstraction at a time, there is no default or ambient authority to steal any undetected advantage.

The PP250 provided 12 transparent capability context registers, as shown in Figure 12. Eight were program-controlled by the functional frame, and four were reserved for specific Church Instructions to unfold the frameworks of a λ-calculus Namespace, the active Thread, an Interrupt Thread, and an Exception Handler (Recovery) Namespace.

The context registers to avoid unfair privileges while sensing all mistakes and detecting attacks in advance, preventing digital harm. The transparent context control registers work the same way for every capability in a computation frame. The handful of Church Instructions and a few context registers cope with one mathematical expression at a time. More symbols signal a poor decomposition.

CR-ID	Function	Microcode λ-calculus	Program Control
0		Load Capability	
1	Five active λ-calculusPro-gram controlled variables		
2			Eight Program accessible Capability Context Registers usingLoad Capability
3			
4			
5	Thread Event Variables		
6	DNA String C-List	Call & Return	
7	Function Context		
8	Thread Context	Change Process	
9	Interrupt Thread	Interrupt	Hidden Church-Turing Instruction Mechanisms
10	Active Namespace	Switch Namespace	
11	Restart Namespace	Error Detected	

FIGURE 25: PP250 CAPABILITY CONTEXT REGISTERS

THE CHURCH INSTRUCTIONS

The Church Instructions add the laws of λ-calculus symbolic addressing to encapsulate the binary commands of Alan Turing. Turing-Commands are now bound to mathematical symbols instead of static linear memory that enforce rules of symbolic binding ranging from a single symbol to covering a whole Thread or a Namespace. The Church Instructions control the movement through the woven mathematical symbols of a directional DNA hierarchy.

Adding the Church Instructions delivers programmatic control over the execution scope, see Figure 25. Digital boundary walls and a digital Ishtar Gate prevent the imperative Turing- Commands from interfering with another domain. Instead of using shared memory, instructions are always bound to capability symbols, and the DNA always groups these symbols into functional nodes.

One capability context register represents one symbol at a time, and instructions are constrained to the geometric digital scope of the symbol as translated by one Capability Key. The Capability Keys are grouped into nodes; DNA strings define the nodes in a directional DNA hierarchy. The EnterOnly Access Right insulates each node, a digital security system that blocks machine instructions from invalidly using the other abstractions linked in a DNA chain.

This directional DNA hierarchy as a framework is a functional structure, a model of the symbols in an expression written on an endless scientific canvas of virtual chalkboards—the machinery of an infinite universal model of woven computations, a Dream Machine for the Enchantress of Numbers.

Each symbol is bound on demand to a Church Instruction's current computation frame. The capability context register is an authorised door to an encapsulated resource. The calculation in Turing's α-machine is a DNA frame within a transactional Thread. Each frame has a DNA string for one chalkboard expression.

An external event creates a transactional Thread. The Thread evaluates the event, which carries a set of event variables through a mathematical chain of related function abstractions. The chain is the directional DNA hierarchy organised to assess the expressions on the chalkboard. The expression-tight computation is validated and verified instruction by instruction. The result is the *Infallible Automation* of the expressions written on the chalkboard that compute one event at a time.

The framework flawlessly hardens and perfectly implements the individual expressions taken in turn from the virtual chalkboard of the namespace. Its deep-and-detailed level of protection creates lean, fast, fail-safe, and future-safe software. Patching is unnecessary because error recovery is automatic, built into the machine when Turing's α-machine is encapsulated by λ-calculus using Capability Limited Addressing.

A parallel hardware pipeline performs the checking overhead without software baggage or instruction delay. Digital boundaries are sustained with the type-controlled access rights that distinguish the core concepts of λ-calculus from attack. Using Capability Keys as parameters between software modules makes the parameters fraud and forgery-resistant, preventing Confused Deputy attacks that debilitate General-Purpose Computer Science. The dangerous clickjack attacks that drive 90% of undetected cybercrimes are detected and prevented. When using immutable Capability Keys as type-safe λ-calculus variables, each variable is validated and verified as authentic in type and form.

Protected by the guard rails of Capability Limited Addressing, programs need minimum code and achieve maximum speed. The performance is further enhanced using anonymous λ-calculus variables and parallel Threads. As a result, efficiency and productivity increase, counterbalanced by the checks enforced by Capability

Limited Addressing and the λ-calculus meta- machine in the basement of the Church-Turing Machine's technology stack. The two counterbalanced designs achieve the highest computational quality for the minimum code. Thus, the highest speed result was achieved with the added advantages of machine-level parallel processing and machine-level anonymous variables.

The Church Instructions create a *'Typed Controlled, Networked Software Machine.'* The context registers enhance algebra, geometry, and precision in machine instruction, adding multi-dimensional strength to Turing's basic logic. The mathematically balanced Church and the Turing machines work together to isolate interfering forces that attempt to break out from the specified activities on the chalkboard. The only code required for any implementation is the minimum functional solution, and the code reuse of the function abstractions further improves software efficiency, quality, and productivity.

The six Church Instructions invented for PP250 are listed in Figure 28. These instructions constrain the scope of Turing's α- machine as a functional execution. Frame by frame, Church- Instruction by Church-Instruction, the functional symbols unfold the digital form of the function abstraction. The digital form is the computational context. It is nonlinear and no longer trapped in shared memory like a general-purpose computer. Indeed, it has a multi-dimensional global scope. The twelve context-registers set the scope of frame registers that cache the digital geometry in the set of twelve Capability Keys (CR [0] to CR[11]), shown in Figure 25.

Transparent cybersecurity through Capability Keys puts users in control of policy decisions over their confidential data. Web Browsers that roam the Internet thrive on downloaded data. In contact with unknown servers, malware downloads make browsers a most challenging design in a General-Purpose Computer. However, in a Church-Turing Machine, downloads are encapsulated as *EnterOnly* frames, protected function abstractions. It is a safe way to contain any digital download. As a dynamic assembly, any λ-calculus violation is a sign of an attack, and a Church-Turing Machine responds immediately and instinctively.

The full range of programmed control over security policy is achieved by the handful of Church Instructions that unfold the hidden springs of nature. It starts with a simple Load-Capability instruction to change the scope of one context register $CR[x]$, where 'x' for PP250 is a 3-bit register ID field from capability register zero to seven. At the other extreme, the Thread context is changed by a Church Instruction that updates every context register (perhaps), including the Namespace. Another Church Instruction is a spontaneous action to an error. It 'Switches' to the recovery Namespace and activates a default Thread of that namespace.

The PP250 cached the Thread in CR [8] and the Namespace in CR[10], while CR[9] is an interrupt Thread used for Interrupt conditions. Also, the accessible program context registers CR [0] to CR[7], including CR[6] to hold the nodal C-List, the *'DNA string'* as a navigator co-pilot and CR[7] to define the *'cockpit-seat,'* as the updated tape-reader of Turing's α-machine.

The remaining context registers CR [0] to CR[5] are used to build results and pass λ-calculus variables between two *EnterOnly* frames executed by the same Thread. By convention, the event variables were found as a function abstraction in CR [5].

The traditional binary data manipulated by a Turing- Command is encapsulated by the context of a mathematical symbol in a context register CR[x]. The binary data is an offset from the instruction constrained by the address range obtained from the Capability Key. Turing's binary data and Church's λ- calculus Capability Keys are stored in memory, but the two machine types are separated.

A Capability Key is not binary data, and if referenced by a Turing-Instruction with an incorrect register ID, an error is thrown. The reverse is equally true. When the Error is thrown, the Thread aborts and the self-preservation instinct built into the Church-Turing Machine takes over. These symmetrical rules protect the immutable Capability Keys and prevent software fraud and forgery attacks.

Capability Instructions		Source	Cached Thread
LC	**Load** Key from a C-List[c].offset	CR[c].offset	→ CR[d], Key[8].d
SC	**Save** Key to a C-list [d].offset	Key[c]	→ CR[c].offset

CAL	**Call** a Function Abstraction froma C-List[c] and function at offset	CR[c].offset CR[c] CR[8].Push	→ CR[7] → CR[6] Instruction Offset
RTN	**Return** to Prior Thread Contextat 'popped' instruction offset	CR[8].Pop	← CR[7] ← CR[6] ← Instruction Offset
CP	**Change** Process to another (suspended) Thread	CR[c].offset	→ CR[8] For x =8 to x=0 ← CR[x]
SN	**Switch** λ-calculus Namespace and select a Thread	CR[c].Switch	→ CR[10], then Change Process

---

FIGURE 28: THE CHURCH INSTRUCTIONS OF THE PP250

Aborting the Thread guarantees that any malware or hack is also aborted. This extreme action prevents any harm, damage, or digital theft to all symbols in the Namespace. The Church Instructions enforce fraud and forgery-resistant boundaries by using Capability Keys as variables in function calls. At the same time, a passive Capability Key extends the reach of computation dynamically across Cyberspace. The algebra of a Capability Key is universal, and the digital geometry of Capability Limited Addressing is open-ended, providing software integrity through functional boundaries in each Church-Turing Machine.

Capability Keys mould software as sub-atomic, atomic, molecular, and cellular organs in a Namespace as a software species, implemented and protected as named functional classes of digital objects. From itemised sub-atomic media to function abstractions to the application as a living set of transactions, the reproducible software species is protected in a data-tight, function-tight C-List obtained from the directional DNA hierarchy. Both accidental errors and deliberately targeted human attacks are detected and stopped when in contact.

THE MAGIC OF VARIABLES

A s a part of this protection, the sub-atomic binary media is limited to *Read, Write,* and *eXecute* binary access rights, while C- List offers *Load, Save,* and *Enter* access rights. The type of limited separation protects the software from manipulation. Structured by immutable Capability Keys as digital tokens of authority applied as the foundation layer, computer science is fail-safe, the prerequisite of flawless mathematics. The λ-calculus meta- machine operates as clockwork rules that prevent corruption, including attempted fraud, forgery of Capability Keys, and typed and scoped 360° of geometric boundary protection.

Only Church Instructions can access C-Lists, and only Turing-Commands can access binary data. Furthermore, the Capability Keys in an *EnterOnly* C-List are unreachable to any request from outside the private scope of the unit of object-oriented machine code. Access to these C-Lists is locked and hidden within the gated walls of each function abstraction as a software black box.

A nodal, object-oriented list of Capability Keys is only accessible after transferring control to the new frame to hide the implementation details to a foreign C-List. The microcode of the frame-changing Church Instruction prevents privilege escalation. It is not under software control.

Each Church Instruction performs a specialised role that unfolds the new computational execution framework. At the same time, the formality of the directional DNA hierarchy prevents outside attacks

KENNETH JAMES HAMER-HODGES

KENNETH JAMES HAMER-HODGES

by any form of malware or user hacking. The architecture of a namespace defines a living species of software rigorously enforced as an atomic, sub-atomic, and chained molecular structure. Each object has a defined and type- constrained digital purpose.

Consider, for example, the human hand. It is a multi-purpose functional organ of the human body. Each of the joints and movements allows limited functions within an overall scheme. The scheme is complex, but at each joint, the rules are simple, and the boundaries on movement can be locally enforced. The limitations are various but superficial and easily enforced without limiting the overall functional capability.

Indeed, the functions reinforce each other, allowing full rotation and detailed articulation, like opposing the thumb with the fingers for humans. All this is defined by the directional DNA hierarchy using simple atomic function abstractions.

To stress this point, it is not the chemistry of the hand that defines the species as a Human or a Chimpanzee; the atoms, genes, cells, and organs are almost identical. It is the structure that makes them different, and the DNA of the species defines the structure. So, it is with the DNA of a λ-calculus software machine as a functional namespace in a Church-Turing Machine. The directional DNA hierarchy is both a formal and a functional level of programming that is missing from a General-Purpose Computer.

Errors confined to the typed digital boundaries will only spread if ignored. A Church-Turing Machine takes immediate action. The defended borders guarded by Capability Limited Addressing prevent errors and stop them from spreading. Any attempt to crack this fail-safe system is detected on contact, and the instinctive reaction is immediate. Detection takes place at functional boundaries in advance of harm or digital damage. These rigorously enforced rules have the mathematical precision envisioned by Ada Lovelace as the magic of numbers.

A clockwork λ-calculus meta-machine scientifically shapes Capability Limited Addressing to encapsulate the type of limited λ-calculus variables. Errors are detected variable by variable in a fail-safe way. The λ-calculus meta-machine applies the laws of the universal model of computation as the science behind every action. Worldwide, over half the cost of software in development and ownership is spent on failing forms of ineffective cybersecurity. Reversing von Naumann's assumption more than doubles software productivity in every measure. It will double the speed of progress in the 21st century and simultaneously make the electronic village future safe. The software is purely functional, like Ada's first algorithm, the Abacus and the Slide Rule. It will last forever with minimal maintenance.

Without this change, the costs, overheads, and recurring damage to cybersecurity and cyber society will only grow. At some point, servicing the harm caused by flawed General- Purpose Computer Science exceeds the value of the flawed software, leaving nothing for progress and improvement. At that point, democratic society stalls as digital dictators take over.

THE LANGUAGE OF ABSTRACTION

Anetwork of capability-linked function abstractions purges the functional discontinuities of General-Purpose Computer Science. The syntax of high-level abstraction can be written on a chalkboard. It can solve any scientific problem known to humanity. The solution exactly matches Ada's dream, through Babbage's *Infallible Automation* modernised as Alonzo Church's universal model of computation. The only constraint prevents duplicate names in a namespace. Any requests to load or assemble a Variable with a name already assigned must be checked and authorised as the latest version of an old name and not a mistake. In a telecommunications Namespace, as applied to the PP250, there is only one '*PP250. Connect(a,b)*' abstraction. Other service providers, such as Google, can offer competitive solutions like *Google.Voice.Connect(a,b).*' The names *Google* and *Voice* are part of a directional DNA hierarchy of Namespace objects found in cyberspace. Each case connects the same or alternative devices based on the function abstractions to the individual object-oriented, machine-code device drivers. Industry standards for global Capability Keys as tokens have already progressed to the point where the directional DNA hierarchy of a software-defined λ-calculus Namespace can be integrated with tokens of any Church-Turing Machine as endless chains of secure Capability Keys.

In this context, a network of virtual chalkboards defines the software in computer science as a flawless, fail-safe, global machine. A software machine from multiple competitive suppliers where the citizens own their data and service vendors provide the function abstractions. Computational chains that can safely link various vendors, all age groups, and any interest sets in any culture. Each citizen uses their preferred security policy to secure confidential data, all guaranteed by the exact science of λ-calculus symbols first learned from a schoolroom chalkboard. The scientifically defined symbols and expressions serve each citizen equally throughout their cyber-secure functional interactions, crosschecked like a student at every level of global computation.

With a Church-Turing Machine network architecture, each child at school establishes a private, cyber-secure Namespace for global interactions under parental control constrained by tangible Capability Keys. Its universal cyberspace resists all forms of interference. It is a framework for individual control without dictatorial defaults or unfair outside authorities. As taught on the chalkboard, science replaces General-Purpose Computer Science's opaque, incomprehensible, and unstable concoctions. Pure computer science, the backbone of the 21st century and beyond, can be introduced to schoolchildren as they embark on life when they first learn to add and subtract.

The cyber-dependent society of the 21st century and beyond demands more than General-Purpose Computers can offer. The opaque and corrupt hallmarks of General-Purpose Computer Science and the undetected crimes against humanity that are antithetical to civilisation must evaporate. Universal computer science based on the Church-Turing Thesis engineered to serve citizens and communities can be taught to all at school for a small effort and huge advantages. Only the science of the Church- Turing Thesis can faithfully serve science and society in this uncertain and uncharted future.

CHAPTER 14

Colonizing Cyberspace

*Commerce with all nations, alliance with
none, should be our motto.*
Thomas Jefferson

The colonization of cyberspace is as big an adventure as the colonization of North America. The early settlements are primitive and vulnerable. The settlers cling to the shoreline of the pre-electronic age. The effort to survive is all-consuming. The deconstruction of physical boundaries by digital convergence leaves the settlers unguarded and exposed. Survival cannot be guaranteed. Security and survival tilt uphill into the heartland of an unknown continent. The early settlers faced constant attacks from enemies, criminals, and remote dictators who owned the ground.

Computer science discovered a new digitally converged world that unavoidably leads to a new understanding of freedom, equality, and justice. One day, the world will be governed through, but not governed by, digitally converged cyberspace. The flawed technology of General-Purpose Computer Science will be replaced with a scientific alternative that the citizens of this information age control. Digital law and order, civil freedom and justice, individual privacy and independence will characterise the foundations of a survivable cyber society as humanity occupies this new world.

Cybersociety will never endure or accept the loss of law and order in General-Purpose Computer Science. Either in advance or because of severe catastrophes, action will be taken to civilise cyberspace. Law and order will be brought to the wild digital frontier where software meets hardware. The colonization of cyberspace must be civilised. Universally fair laws of mathematical science will prevent crimes. Justice will be established that applies equally to everyone. It may take centuries as it did with the settlement of America. If so, there will be tragedies along the way. However, the solution is clear, and the tragedies could be avoided by early government action.

Digital convergence bypasses the physical world. The traditional barriers of nature and life are virtualised by software in cyberspace. The established institutions of a civilised democracy are reconstructed through virtual reality. Its result cannot favour the skilled over the innocent, the privileged and the powerful, the corrupt and the uncivilised, or the artificial over the real. It is reasonable to perceive a Deep Fake as a synthetic, virtual reality candidate standing for an election, fabricated holographically by virtual reality and represented by digital convergence but with questionable ethics and no morality.

AI is changing the game, yet again, and ending the life cycle of General-Purpose Computer Science. Advanced software and backward hardware only increase and further complicate the problems of colonizing cyberspace. Unelected administrators and branded cyber monopolies go hand in hand with hidden criminals and unseen enemies who distort justice through undetected malware and hacking to gain some advantage. As digital crimes increase, insecurity grows, and trust in government is undermined.

Insecure, unstable, and dictatorial computer science is replacing the founding vision of American freedom, constitutional democracy, and government by and for the people. The unregulated General-Purpose Computer is virtualizing and, simultaneously, breaking every law, tradition, and precedence that defines the American way of life. Life in the 21st century cannot be run by the unfair privileges ruled by the unknown, unelected forces of cyberspace that now include Artificially Intelligent Malware.

The General-Purpose Computer is digitally unfair and numerically unsafe. Self-centred powers and unelected dictators expand the criminal class and empower open warfare in cyberspace. The attacks are silently launched from international mobile hubs that remain undetected. The forced changes in society are surprising, dramatic, and devious. Corruption is destructive, instigated by the undetected, crafted attacks by hidden, unseen foreign armies. Cybercrime and cyberwar have become existential threats to the lives of individuals and nations sharing the electronic global village.

Interestingly, colonizing cyberspace is a double-edged sword. This sword cuts both ways, and it cannot be ignored. When America was colonised, only the settlers faced hardship. With cyberspace, everyone is a settler because cyberspace is the new reality of life. It cannot be ignored. It only takes one major cyber catastrophe to destroy trust and stall a nation. The unfair privileges created an unelected ruling class of cyber criminals, and specialised experts distributed unevenly. The lack of confidence in national security is self-evident. Citizens and society are suffering from unjust digital laws and unequal justice, unfair elections, and digital slavery is the end game. Unfair cyberspace degrades life to the lowest common denominator of interconnected societies. Eventually, digital entropy is lost, and the ability for cyberspace to work will grind to a halt.

The digital clock is turning life back to a prior time, a middle era when power and privilege were unjust and inherited or regal. At the same time, citizens are branded serfs, digitally enslaved people, or monopolistic property. Transparent law enforcement in cyberspace is vital to protect the poor, the weak, and the sick by levelling the playing field for all and preventing hidden crimes and undetected enemy encroachment.

While good things get better, bad things only get worse, and things are getting worse. The arrival of artificially intelligent malware will increase the rate of successful attacks, and industrial society will stall. Cyber insecurity is the root cause of industrial woes; the cost to sustain this failed approach prevents sound investments and safe solutions. The flawed assumption of blind trust in shared memory and perfect programs turned General-Purpose Computers into the enemy of industrial society.

Digital convergence is the tool of choice for international criminals and enemies of the state. The unregulated powers in cyberspace silently attack civilised society. Regulation by the government that does not address the root cause will only place more restrictions on innovation and further damage progress. Colonising cyberspace requires the same attention to survival standards as colonising a space station, the Moon or Mars. Minimum standards for computer science are required. The root cause of malware and hacking must be fixed before the symptoms of insecurity can be regulated. Confidential data must belong to the individual and be placed in the hands of the citizens to manage their digital property safety if cybersociety is to remain a democratic government of the people, by the people and for the people. The threat of a dictatorial breakout by artificially intelligent malware must be prevented.

The workable solution changes the nature of digital convergence from branded dictatorships to private software machines. To safely colonise cyberspace, each citizen needs a private namespace as a data-tight cyberspace suit. Using a Church-Turing Machine, the data-and-function-tight suit is a λ- calculus Namespace reinforced by capability-limited computations. As proposed by the λ-calculus, a digital version of Babbage's Infallible Automation will safely extend individuality into cyberspace. It is Industrial Strength Computer Science, where cybersecurity is personal and private. Digital survival is individual, and cyber democracy remains a government by and for the people.

A λ-calculus Namespace in Industrial Strength Computer Science is guaranteed safe and secure for civilian use. It is faithful, fail-safe, and future-safe—the media-tight servant of humanity. As servants, computers are controlled by individuals, industries, citizens, communities, and nations as the actual owners instead of branded suppliers selling dictatorial monopolies, unelected third-party administrators and unsafe, centralised operating systems.

Industrial Strength Computer Science is the future-safe and fail-safe, obedient servant of individuals and society. It is the digital manifestation of Babbage's hallmark and Ada's dream, as found in the Church-Turing Thesis. Infallible automation is guaranteed by the balanced digital technologies of the Church- Turing Machine as an architecture for endless computer science. The converged digital societies of the 21st century must use science to avoid a dark digital age of branded dictatorships and decay.

THE NATURAL LANGUAGE

Mathematics is the universal scientific language behind every law in the universe. Civilisation began with the universal language of arithmetic captured by the Abacus, and it will culminate with Ada Lovelace's endless mathematical dream of Babbage's *Infallible Automation*. Mathematics regulates and explains every hidden force and every unexpected or spontaneous reaction in Mother Nature. It governs the regular movement of both heavenly bodies and earthly living species. They all survive in a hostile, competitive environment according to the laws of mathematics that govern every Rose and every Abacus. Mathematics is Bronowski's hidden spring of nature—a spring of the mathematical function abstractions in the Church- Turing Thesis and a Church-Turing Machine.

In a Church-Turing Machine, computer software is trusted because Namespace by Namespace, abstraction by abstraction, and function by function software serve science and society. When defined by a λ-calculus meta-machine using Capability Limited Addressing, the symbols are automatically safe as trusted digital components. As such, the abstract software is translated into dependable digital media that resists the inhuman hostility in a global cyber society.

The λ-calculus drive the Church-Turing Thesis as the universal model of computation as the scientifically complete alternative to a General-Purpose Computer. The machine is without cybercrime, cyberwar, and the unfair privileges used by undetected hacks and silent malware. A level playing field is a scientifically complete platform for civilised progress into and beyond the 21st century. The unique characteristic of a network of Church-Turing Machines is the universal enforcement of media-tight mathematical integrity. Each λ-calculus namespace is a data-tight cyberspace suit for the citizens to colonise the future. As a scientific solution, there are no assumptions, and like Babbage's Difference Engine, there are no human practices to corrupt results. It is a digital mirror, a mathematical utility that executes directly from a virtual chalkboard of scientific expressions.

Like a Rose or a Rainbow, a flawless servant, a fail-safe machine, and a future-safe computer, these trusted computers are always faithful to nature, mathematics, functionality and humanity as a predictable, reliable software machine. The wild forces created by malware, inspired by humanity or AI, are isolated, contained and locked out of the cockpit. If a booby trap explodes, it remains fail-safe, and this never changes because the pure design is scientific and the servant of society.

Even Artificially Intelligent Malware is inspired by humans. Even when AI reaches a superhuman level, Artificially Intelligent Malware is contained and constrained to the criminal's namespace and cannot move beyond to attack others. In a Church-Turing Machine, software breakout is prevented.

Wild rivers of malware and sabotage caused by hackers leading to unregulated digital floods no longer destroy communities. They are digitally constrained to a Namespace using the force of nature used throughout history. They can all be safely engineered through scientifically applied technology. Malware and hacking are distorted forms of mathematics that can, if unchecked, destroy a nation. However, the laws of λ-calculus regulate mathematics. The universal model of computation governs everything in nature, including computer science.

Engineering software through the λ-calculus perfectly serves both society and science at the same time. Mathematics allows engineers to span ravines with roads and build rockets that reach the moon and return safely. Software is the same; it must be engineered to survive and pass every test of time in a hostile environment where danger is natural, even digitally supernatural AI. It is how to serve humanity beyond the 21st century for the dreams and vision of the founders of civilisations to survive in cyberspace.

AN ENCHANTED FORMULA

An Abacus serves society as a trusted machine for eternity. It is limited to arithmetic and requires memorisation of the algorithm, but otherwise, the machine operates flawlessly. The enchanted formula of flawless automation started with the Abacus to enhance human power and limit human error. The magical formula took a step forward with the Slide Rule. Now, the mathematics moved from the head to the eye as the enchanted formula of a logarithmic scale embedded itself into the machine as a mathematical spring.

Bronowski's Rose, a snowflake, a rainbow and a honeycomb epitomise the enchanted formula of Infallible Automation. Life in Mother Nature's universe goes all the way to automate the removal of defects. Babbage achieved flawless engineered excellence with his prototype, the Difference Engine, and anticipated the λ-calculus with his proposal for the Analytical Engine.

Civilisations expect this timeless, flawless service in every engineered undertaking. The errors in science and engineered technology limit the progress of society and civilisation. Since the Abacus, scientific excellence has moved hand in hand with civilized progress. However, the dark side of digital convergence is unscientific and unexpectedly enlarges the threats to human life. The machinery of a digital computer that is fit for the 21st century requires software to pass the test of time as an engineered, faithful, functional servant.

As a faithful servant, running software-defined machines that reach worldwide works reliably for future generations, as the Abacus and the Slide Rule of the past. The unfair despotic privileges of General-Purpose Computers endanger life in the electronic village. All unfair privileges that distort and kill the mathematics of life-supporting software must be replaced. The essential first step is adding the λ-calculus meta-machine to encapsulate the binary computer and regulate software as a coherent functional machine.

Computing mathematics symbolically with locally substituted variables moves the universal model of computation as an enchanted formula into an everlasting future. The encapsulated symbols as function abstractions and data variables inherit the fail-safe automation of a capability-protected binary computer as the digital gene in a Church-Turing Machine.

Notably, the Church-Turing Machine is scientifically faithful to the expressions written and understood from a schoolroom chalkboard, exactly as Ada Lovelace wrote her first program two centuries ago. These future-safe, fail-safe advantages a statically bound General-Purpose Computer cannot match.

A Capability Key to a λ-calculus abstraction is an immutable name to a digital implementation; it is not a physical address in shared memory. The digital gene in a Church-Turing Machine only recognises the mathematical names, leaving the implementation details to separate function abstractions in the λ-calculus meta-machine to work out. These digital names define an immutable, trusted currency of computation and elevate the enchanted formula as a cyber suit to every individual settler in cyberspace.

Unlike corruptible binary data, the citizens of cyberspace use immutable Capability Keys that resist corruption. Using easily forged binary data leads to many unresolved, *'confused deputy'* attacks that beset innocent citizens. The enchanted formula of a λ-calculus namespace immunises innocent citizens from unseen, silent, digital manipulation. A Deep Fake newsreader on TV and a Deep Fake election winner are seminal examples easily detected by the directional DNA hierarchy of the Church-Turing Thesis, the universal model of computation and capability-limited addressing.

A cyber-suit is equipped to keep citizens safe in cyberspace. The cyber-suit is a λ-calculus Namespace for the individual. It is the incarnation of their private digital shadow, using their private DNA as their ever-present capability secure desktop. As desktop tokens, Capability Keys unfold the abstractions of software as protected digital structures, their private Babylonian city-state in cyberspace, guarded and protected from malware, hackers, and superhuman attacks organised by Artificially Intelligent Malware.

The personal λ-calculus Namespace extends individuality into cyberspace driven symbolically by heavily protected, private Capability Keys. Every action taken by the desktop is validated and verified on the spot. The bound result has digital integrity, guaranteed by the Namespace cybersuit. Each cyber-suit is dynamically guarded day and night. Immutable keys signify access rights to the Ishtar Gates of these digitally walled city- states in cyberspace.

Capability keys, alternative machine instructions, and functional Church instructions, as needed to secure cyberspace, are built into a cybersuit's encapsulated workings. Turing's α- machine only computes exposed binary data while a λ-calculus meta-machine navigates cyberspace safely, exploring the directional DNA hierarchy chosen by a user. The tokens of power and authority are iconic Capability Keys that bring Capability Limited Addressing to the visual surface of cyberspace.

The iconic symbols are the immutable tokens of the functionality implemented by the private cyber suit as a named λ-calculus assembly belonging to a settler in cyberspace. At every level, Capability Keys express the enchanted formula of *Infallible Automation* essential to implementing power in a namespace.

In a cyber-suit, a personal software machine defends vulnerable, innocent citizens. A capability-based cyber machine is a secure digital framework that extends human life into the 21st-century electronic village. Depending upon progress, the ultimate objective is trusted information exchange. All data imported into the Capability-empowered cyber suit is validated and verified to prevent Deep Fakes and False News. Capability Keys track all information back to the source to identify all threats and corruption.

Capability Keys make all the difference to the timeless implementation of faithful mathematical machines and the colonization of cyberspace. These software machines are flawless cyber machines, shielded and protected by the language of λ-calculus, to defend the integrity of fail-safe, future-safe digital media. The symbols of the segmented digital implementations are protected from harm caused by unchecked Turing-Commands on the wild frontier of computer science.

THE CYBER SUIT

The security of the settlers in cyberspace depends on restructuring computer science. Every assumption of General-Purpose Computer Science must be turned upside down and inside out. It starts by reversing von Neumann's foolish assumption of blind trust in software. Alonzo Church had already proved there is more to computer science than just Turing's α- machine, but von Neumann ignored the λ-calculus and overstretched Turing's α-machine.

The Church-Turing Thesis hides a second machine type. It is a λ-calculus meta-machine that scales the Turing α-machine without overstretched commands or unfair default privileges. Two kinds of digital information exist in mathematical cyberspace: binary information and immutable typed capability data. The capability data is the meta-data for software as a protected digital machine. In a Church-Turing Machine, the two data types are distinct, never to be confused. In the universal model of computation, only the laws of λ-calculus apply.

This meta machine is an atomic cyber suit for dangerously wild binary computers. The cyber suit is reinforced by the meta-data of chainmail that seals the data-tight, function-tight, and media-tight boundaries of the functional mathematics in cyberspace. Meta-data is fundamental to scalable cybersecurity and the reliable evolution of software. Before the first distributed computers were invented for telecommunications, the λ-calculus as a 'Church Machine' remained hidden, but it is vital for secure, scalable software.

It takes pride in being at the base of the technology stack in a Church-Turing Machine. The machine computes symbolically as protected function abstractions that pass the test of time and reach worldwide. When the λ-calculus cyber suit is built into the foundations of computer science as a Church-Turing Machine, software as a program is encapsulated and governed by a directional DNA hierarchy of mathematics. A λ-calculus namespace is a stable platform for an evolving digital species that can roam networked cyberspace and survive. Each software species is encapsulated by a directional DNA hierarchy structured by relationships.

The Capability Keys of Capability Limited Addressing form λ-calculus relationships as symbols defining individual machine types in the universal model of computation. Because the λ- calculus meta-machine is transparent in a Church-Turing Machine, it is also software-controlled, programmed by functional Church Instructions. The Church Instructions unfold a just-in-time context of execution for Turing's α-machine.

The nodes in the directional DNA hierarchy of Capability Keys are grouped as DNA strings in C-Lists. It builds a secure framework for the application based on the universal model of computation. When Church Instructions unlock Capability Keys, the digital petals of function abstractions are unfolded by Church Instructions as a blueprint of functionality. The functionality is constrained, and at the same time, the protected DNA hierarchy guides the infallibly automated mathematics.

This digitally woven fabric is as natural and beautiful as a flower opening and closing by following the sun. The directional DNA hierarchy is a digital mirror on the surface of computer science. The Church Instructions navigate the digital reflection opening and closing the function abstractions in response to events. These events represented by Threads call on function abstractions as endless chains, universal mathematical cyberspace that faithfully serves humanity by serving science in full. It was the vision foreseen by Ada Lovelace as she worked with Charles Babbage on the Analytical Engine.

Thus, Turing's α-machine is the cockpit of a software species; it is a digital gene in a directional DNA hierarchy. The pilot and a co-pilot in a shared algorithm navigate the encapsulated function abstractions, piloting the universal model of computation ruled by the laws of λ-calculus. The algorithm is a hidden spring that unfolds other petals in the framework. The framework defines the context for imperative Turing-Commands to colour and shade the properties of the growing result.

The λ-calculus meta-machine is a computational gearbox with built-in guard rails to navigate between computational frames safely. The gears change the context by unlocking the appropriate access rights to named objects. The Church Instructions follow the chains of Capability Keys as peer-to-peer, local or remote, computations that reach securely, on-demand, right across cyber-space.

The two machines offer two different and complementary forms of machine language. One is the traditional mathematical logic for statically bound procedures used by General-Purpose Computers but bound to symbols instead of static memory, and the other machine language adds functional syntax for high-level λ-calculus expressions created by the symbolism of the immutable Capability Keys.

These Church Instructions become application-oriented statements by referencing the application-oriented names in the directional DNA hierarchy. The example used earlier of the symbol to variables called *Switch, Connect,* and *myNumber* needed by the telecommunication application are good examples. The names establish an application-oriented navigation language built into the directional DNA hierarchy of each λ-calculus Namespace.

It is a high-level, application-oriented language that operates at the machine level. What could work better than that? As discovered when programming the PP250 without a compiler, the object-oriented machine code with the Capability Keys enhances productivity, performance, and the quality of the result. It leads to the view that citizens can use the λ-calculus namespace as a functional, programmable cyber suit. Its protected mathematical environment is needed to prevent the dictatorial monopolies from running the world. Each citizen empowered by mathematics and tangible symbolism will control their destiny in cyberspace.

The code is minimised because this high-level machine language is application-oriented, built from λ-calculus Variables, Abstractions, Functions, and Applications as a digital Namespace. At the same time, the performance is maximised, and the quality improvements include profound cybersecurity.

Indeed, the machine types are the interworking components of a reliable application-oriented machine defined by software but sustained by the laws of λ-calculus. The Church Instructions are the gears in this universal model of computation. For example, for those familiar with the Wolfram Language or other multi-paradigm symbolic languages, all abstractions needed can be built to operate better on a Church-Turing Machine. The user interface for a Namespace is a graphical chalkboard for mathematics and functional activity.

The Church-Turing Machine achieves all this because General-Purpose Computer Science's overheads, threats, and downsides are replaced by pure science and solid technology. Each namespace is the servant of the application-oriented capability keys that provide the user with the exact authority needed for the specific computational activity, and the namespace is the mathematical servant of the user.

MACHINE CONVENTIONS

Like other computers, a Church-Turing Machine has register conventions to pass variables between function abstractions. Each parameter must be checked to avoid the accidental mistake of an undiscovered bug in the code or a deliberate error created as part of a malware attack. Using a General-Purpose Computer is impossible and is the root cause of many Confused Deputy attacks. Such attacks are prevented when parameters are Capability Keys. The parameter passed cannot exceed the authority or scope of the provider.

Parameter passing is simplified by the type and access rights checks performed with every machine instruction. For example, when passing the function abstraction of a decimal integer, it remains self-verified and true-to-form. The type can be easily checked before use, reducing the complexity and size of the codebase.

The Babylonian Golden Rules made the Abacus an everlasting mathematical machine that works for the abstractions of arithmetics. Software best practices are inherited instead of being painfully learned at high recurring costs. It reduces checking overheads in cases, allowing for no additional code.

The register conventions used by PP250 support the parameter passing of both Capability Keys and binary data. An example is shown in Figure 29. In this example, the function is expressed by the names of the Capability Keys *ClassSwitching.FunctionConnect(this,that)*.

The twelve-context registers (CR[0] to CR[11]) are established by executing the Load Capability, Church-instruction. Only the first eight are under local program control, and specific church instructions manage the four specialized λ-calculus capability registers.

Id	Call Convention	Symbolic Name	Return Convention
CR[0]	Key to package	E.g., Connect	Contains Function of Abstract Class
CR[1]	Key to the first Operand	E.g., +11234567890	Returned first Operand (Changed)
CR[2]	Key to the second Operand	E.g., +4407234512345	
CR[3]	Key to the third Operand	E.g., A Flag object	Messaging between distributed Threads
CR[4]			
CR[5]			
CR[6]	DNA Scope C-List	E.g., *EnterOnly* C-List	Push/Pop to DNA C-List
CR[7]	DNA eXecute function	E.g. eXecute code	Push/Pop to eXecute Function
Hidden λ-calculus Registers used by trusted O-O micro-coded Church Instructions			
CR[8]	Thread Context	E.g., MyThread	Only Changed by Church Instruction
CR[9]	Interrupt Thread	Anonymous	Changed by interrupt microcode
CR[10]	Namespace Table	MyNamespace	Changed of λ-calculus violation
CR[11]	ExceptionHandler Namespace	MyExceptions	Takes control after aFault

Traditional Data Registers are shown below.			
D[0]			
D[1]	First Value Operand	Some binary value	
D[2]			
D[3]			
D[4]			
D[5]			
D[6]			
D[7]	Binary return value	0=Success, -1= failed, -2= bad parameter, -3 = bad template, -4= bad type	

FIGURE 29: THE REGISTER CONVENTIONS OF THE PP250

In addition, the typed λ-calculus Variables for the Namespace, the Recovery Namespace, the Thread, and an Interrupt Thread depend on specific Capability Keys only available to particular abstractions.

The eight registers D[0] to D[7] are traditional binary data registers reserved for the binary RISC instructions. PP250s register conventions were established by a code snippet, a macro expansion to abstract the Church Instruction as high-level, application-oriented expressions using locally named Capability Keys. There are no additional concerns to intrude on the real programming context. No operating System Calls are needed, no virtual-machine boundaries encroach, and no heavyweight virtual-machine processes to consider while hiding all details within the function abstractions as defended computational frames. Each frame is a separate DNA string as a node in the larger directional DNA hierarchy of the Namespace that keeps the software on mathematical rails.

DATA OWNERSHIP

Church-Turing Machine substitutes the λ-calculus Variables with local implementations. The chains of Function Abstractions used by a Thread are unique to the context of the Thread or the Namespace. The computations in a λ-calculus Namespace are owned and governed by the policy of the owner of the Capability Keys as mathematical symbols. Thus, a Namespace as an execution environment for Threaded software differs significantly from a Virtual Machine in the following important ways:

- No shared physical memory exists, and no default privileges exist. The only shared ambient authority is the machine registers of the encapsulated binary instructions in a single computation thread. Computations are contained and defined by the in-scope Capability Keys.

- Threads compute within the scope of the delegated capability keys that are reachable by a binary computer as the computational gene in the universal model of computation.

- The Capability Keys are immune to fraud and forgery. They use immutable tokens to guarantee the laws of λ- calculus enforced by Capability Limited Address space technology. They are tickets to digital power and authority as ordained by the programmer and resource owner.

- The policy of substituted implementations for each λ- calculus variable is privately defined by the user or a community of users sharing the security policy of a private or community Namespace.

- The function abstraction can be purchased or leased and downloaded from network service providers to users through networked Capability Keys, keeping the data private to the user under the policy control of the Namespace owner.

- The user can subscribe to different functional abstractions and control all the computations that co- exist in the user's Namespace.

- The computational Threads have no statically bound resources; immutable Capability Keys dynamically bind everything. It is this DNA that structures the security policy of the Namespace.

- A Thread is characterised by a λ-calculus Application where the Namespace owner defines the event variables, and the function abstraction is navigated by the directional DNA hierarchy as defined by the services chosen by the user.

- At all times, the user data and the computational data created by the λ-calculus Namespace remain under the ownership of the Namespace. No local or remote capability connections can be added to the Namespace without the owner's approval.

- All actions, including data sharing, are context-limited to the λ-calculus Namespace. The Namespace owner controls any networked communication. The scope of a functional chain is approved by the user, not a service provider.

- The Church Instructions that navigate the directional DNA hierarchy as protected computational frames also remain under the Namespace context, and any remote communications are subject to the Namespace policy rules.

- As computational genes, threads carry private variables to function abstractions constrained by the data owner, not the service provider.

Unlike a virtual machine statically defined by a frozen binary compilation, a dynamic computational Thread uses the universal model of computation as a dynamic digital survival machine. The Thread has a cockpit for a function abstraction with a seat for the pilot program for the Turing Commands and a co-pilot for the Church Instructions as the DNA navigator. Both have automatic self-protection mechanisms. The universal model of computation might fail for one of the λ-calculus errors, but the crew and the passengers reach the ground safely.

The Thread in CR[8], the navigator in context registers CR[6], and the pilot in CR[7] and the other machine variables are initialised by a function abstraction representing events in the computation. The starting node traces a path through the directional DNA hierarchy by carrying the event variables to the function abstraction precisely as Alonzo Church defined a λ- calculus Application. The navigator guides the pilot through the directional flight plan using the DNA hierarchy as a map in CR[6]. Starting from the event, the Thread colours and shades the result using the event variables obtained from CR[5].

Each Thread caches its private Variables in CR[5] for easy access to calculate a path through the directional DNA hierarchy. Its path can split into parallel computations, either local or remote, synchronised by Dijkstra Flag abstractions, one of the event variables held in CR[5]. Because everything is referenced symbolically and bound dynamically, the mathematical logic is decoupled from the resource limitations of the physical implementation.

When a connected device signals an event, the driver activates a parallel thread to process the event variables. As an inherited parent of the root function abstraction, a Thread manager can run, pause, stop, and restart a Thread or a set of Threads.

The Thread manager allows a review of register values and Capability Keys during an approved debugging session. The genuinely significant point in this secure transactional architecture is the separation of data from the functions. It enables various execution strategies for the user to control; the Church-Turing Machine architecture is designed to solve user problems.

Nothing takes place without a Capability Key provided by an event and a Thread supplied by the user. Each thread is a computational pump that works for the end-user who owns the Namespace, not for a functional service provider of an unfairly privileged machine.

SOFTWARE RELIABILITY

A stable and tested Namespace is qualified and calibrated to deliver and sustain an agreed level of reliable performance measured in terms of error rate before releasing to the field. Its release defines a generation of stable, functional service measured by mean time between thread errors. The PP250 goal set a nominal rate of less than five errors in 100,000 active thread transactions. The errors are always fail-safe without data loss, corruption, or theft, and speedy return to full service.

The automatic self-healing recovery is built into the verification and validation checks of Capability Limited Addressing integrated with the λ-calculus meta-machine. Capability Limited Addressing protects the software modules, and the qualified service levels are sustained instinctively through *Infallible Automation* without human intervention and loss of digital integrity of information or ownership. Only hardware failures require human intervention, but as a fail-safe solution, a Church-Turing Machine that supports a multi-core configuration used by PP250, one or more cores can be out of service. At the same time, applications continue at a reduced performance rate.

Thread-based software failures are all recoverable. An error-prone function abstraction in a namespace self-identifies and can be switched off, isolated and repaired asynchronously. As a further demonstration of stability and reliability, the PP250 supported concurrent Namespace development on a live system. At worst, an error (also read as an attack by malware or a hacker) impacts only one thread and one event. A massive data breach is impossible because the unfair privileges of a superuser do not exist in a Church-Turing Machine.

Likewise, urgent patching disappears since the recovery mechanisms automatically reestablish the qualified and accepted service level. The system automatically restores an approved level of service quality while measuring and isolating any unstable digital objects, including any downloaded from the web. The recovery process removes the access rights to such modules, marking the object's slot in the Namespace table as out-of-service to prevent further use.

Since Namespace development can take place on a live system, the coordinated and compilation delays that add weeks and months to development schedules are removed. Instead of months between new builds, the turnaround time of a new function abstraction is immediate.

A fix can be reassembled asynchronously if a software class fails and is then downloaded automatically from the cloud during routine service. At the same time, the actual MTBF is continuously monitored for each abstraction while in service. If the MTBF changes, the online feedback identifies the software modules that need improvement in priority order, adding their error statistics. Since the hostility of the environment also evolves as cyberspace evolves and the network grows, feedback sustains stability and allows industrial strength abstractions to improve and pass all the dangerous, uncertain tests of time.

The software errors are all detected by the typed boundary checks of Capability Limited Addressing. A λ-calculus meta- machine violation leads to one of the events listed below, and the Thread immediately aborts the instruction to prevent any digital damage or theft. The recovery Namespace is then activated by reloading the Namespace context and activating the recovery Thread using the allocated Capability Keys and the failing Instruction. The Capability Limited Addressing violations are summarised as follows.

- A Memory Access Violation caused by an address plus data parity or other redundancy checks

- An incorrect address construction that is outside the bounds of the identified context register

- A Capability Key type violation caused by an invalid instruction

- An invalid object Access request or mode violation

- Invalid function abstraction if the 'co-pilot' navigator in CR[6] is not *'EnterOnly.'*

- Invalid program if the 'pilot' program in CR[7] is not *'eXecute.'*

- A stack exception caused by overfill, a CALL beyond the LIFO stack limit size

- A stack exception caused by a RETURN below the LIFO root abstraction

- An exception due to a NULL capability context reference

When an error involves an anonymous Capability Key (passed with no formal name), an octal number is used as a unique identifier to trace the Capability Key in the Namespace table. The Namespace table identified the details of every Capability Key in a Namespace, including any anonymously passed Capability Keys.

DETERMINISTIC GARBAGE COLLECTION

O ne of the unsolved problems in a General-Purpose Computer is memory loss. The shared memory system accentuates this problem in two ways. First, any failure leads to the loss of a global resource that impacts all activity in a virtual machine. Secondly, the shared pages lack memory structure, making garbage collection a traumatic experience. Even if all processing is halted, the memory cannot be easily cleaned. The loss of any computation creates a significant loss to the whole application supported by the virtual machine. Typically, a memory runout leads to an unexpected restart or a blue screen event. These resource problems were unsolved by the WWII misunderstanding of the Church-Turing Thesis.

In a Church-Turing Machine, the Namespace lists the detailed atomic framework of dynamic memory. The Namespace capability table lists all active Capability Keys, including Capability Keys released from use. These Capability Keys identify storage that can be recycled through the memory pool. The PP250 provided a Store Facility Abstraction to manage the allocation of Capability Keys and the typed access rights to corresponding digital media. The function abstractions that exist in every Namespace regulate the private memory resources of the Namespace. Its function abstraction provides several services.

- Request a new segment (either a binary data segment or C-List)

- Reduce access rights (no type of object change allowed, no new rights permitted),

- Change single segment size (max size 196,608 bytes)

- Test Key (size, type, access rights)

- Release a Capability Key (recycle the slot in the Namespace Table and the attached media)

- Create Passive Capability (used as a Credential)

- Create a Literal Capability (22-bit value)

- Garbage collection (neutralise outdated Capability Keys and collect and release lost memory)

When a dynamically allocated segment is no longer needed, it is free and can be explicitly or implicitly released. An implicit release occurs when no Capability Key remains in circulation for the slot identified in the Namespace table. If the slot in the Namespace table identifies a C-List, the release does not apply to the listed Capability Keys since other copies may exist in different C-Lists. The directional DNA hierarchy can include duplications and cross-connects created by dynamic activities. Furthermore, the last copy of a Capability Key may be in the context of a Thread still running on another Church-Turing Machine. Thus, when the Capability Keys are released, copies of the Capability Key may remain and must be removed before the slot in the table can be reused.

Scanning the Namespace Table, the Capability Keys for these redundant entries are found and recycled. A disabled slot traps the out-of-date Capability Keys, and the microcode deletes the Capability Key on the spot. A 'null' access right and a trap condition prevents access to the non-existent object. It prevents any further use.

However, a Namespace slot is only reusable when all copies of the Capability Key disappear. It requires suspending and restarting all threads to clear any released Capability Keys remaining in an unlocked capability context register. By cycling through the C-Lists from the Namespace table, each Capability Key is checked to count the Capability Keys concerning each slot. Unlike a virtual machine, this problem is deterministic in a Church-Turing Machine.

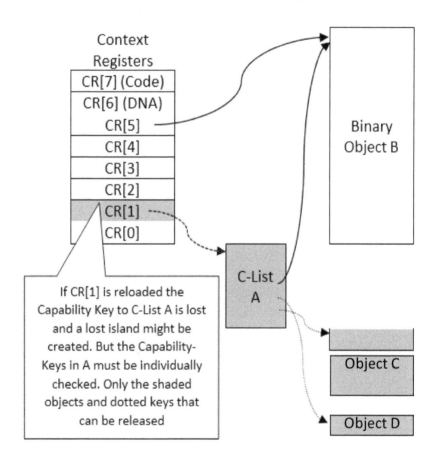

FIGURE 30: THE CAPABILITY STRUCTURE OF DETERMINISTIC GARBAGE COLLECTION

All forms of memory leakage are discovered and recycled over time without halting normal activities. As dynamic events proceed, the last Capability Key to an object, including C-Lists, leaves an island of unreachable digital media. All these islands are found, and the unused objects and slots are returned to the Namespace pool.

Complex examples exist when anonymous C- Lists and anonymous Capability Keys are released. However, the Deterministic Garbage collection resolves every case of complexity to free the slots, the Capability Keys, and the attached memory.

This background program is most efficient when Capability Keys are only held in C-Lists. Efficient garbage collection allows for the natural loss of any Capability Keys and the clean recovery of memory and all physical resources. With a segmented directional DNA hierarchy, recovering these idle slots and attached memory is part of the fault-tolerant system to avoid all blue screen events.

The background Thread cycles through the Namespace Table, counting Capability Keys to set and reset the slot entries and searching for unused slots in the table. At the same time, signals from the Load-Capability microcode confirm the Capability Keys that are in active service. At the end of the cycle, all the garbage locations are identified and recovered. It takes two periods to set and unset the slots in the Namespace table and find the out-of- date Capability Keys, free slots and any attached memory. The Church Instructions capitalise on the microcode to speed deterministic garbage collection as a natural background task.

POLICY CONTROLLED CYBERSECURITY

A ll the technology has been proven to establish cybersecurity control and distribution to programmers and users, not centralized vendors. The migration to a Church-Turing Machine architecture as a faithful networked servant of the user can be achieved as part of the migration to an Internet of Things. The transparent Capability Keys place control of cybersecurity in the hands of society. It is the most valuable advantage of extreme Church-Turing Machines. The Capability Keys are the ultimate control over digital power and program authority that cyber- society citizens can evenly hold to protect their digital privacy.

The Capability Keys elevate cybersecurity to policy-controlled chains of function abstractions, Threads, and confidential data directly controlled by users. The migration changes the nature of cyberspace from a capricious dictator to an obedient servant. A Church-Turing Machine cannot work without the direct order bestowed by a Capability Key since all default powers seized by a Main() program in a General-Purpose Computer are replaced by the λ-calculus meta-machine. When a General-Purpose Computer is successfully attacked and turned into a zombie as part of a 'Bot-net,' the Main() program is the attack target.

No centralised target exists, and no monolithic powers to attack in a Church-Turing Machine. All power and authority are deconstructed and reformed by the laws of λ-calculus. The only power of a Church-Turing Machine is to serve events of limited scope as defined by λ-calculus Variables, Abstractions, Functions, Applications, and Namespace objects that can only be accessed with a Capability Key. Malware and hackers cannot breach these need-to-know, least authority, typed boundaries. Events are processed as λ-calculus Applications, where confidential data is not exposed to outsiders and remains private.

There can be no massive data breach at the machine level because no powers reach beyond a single data instance. Data ownership is private and not centralised as in a shared database. It is transaction-oriented, as in a blockchain. However, blockchains in a General-Purpose Computer remain exposed to successful attacks.

Capability Keys are delegated case by case, thread by thread, and subject by subject. No superuser exists to override the architecture; each application is secured by its own Namespace as an independent city-state in cyberspace. A capability-limited digital cyber suit protects each individual's journey through hostile cyberspace. Each Namespace operates as a cyber suit, a self-contained machine not statically bound to hardware as in a General-Purpose Computer. The user controls the cybersecurity policy in their cyber suit at every step as an independent, secure Namespace.

This user control over cybersecurity is essential to coexist and survive economic, cultural, and political conflicts across the electronic village of the 21st century. The Church-Turing Thesis provides the scientific solution, where everything required by the individual in a cyber-dependent democracy is guaranteed. However, more than this, policy control is in the hands of cyber citizens to keep confidential data private, owned, and managed by a Capability Key.

The Church-Turing Machine is engineered by the laws of λ-calculus to remain steadfast and faithful to mathematics as a predictable, proven science. As Charles Babbage and Ada Lovelace envisioned, computer science is the loyal servant of society's citizens and the symbols in mathematics. It protects the 21st century and beyond from catastrophes. It saves the nation from lost cyberwars and protects society from the punishing digital subjugation of super-intelligent malware. Nothing less is acceptable for the survival of a functioning democracy ruled by the settlers who colonise the digitally converged electronic village of the future. Democracy, freedom, and privacy must last forever and serve as a gift to the ages, just like the Babylonian gift of the Abacus.

CHAPTER 15

The Future of Cyber-Society

*"We live mythically and integrally... However,
continue to think in the old, fragmented space
and time patterns of the pre-electric age."*
Herbert Marshall McLuhan, (1911 – 1980)
*Understanding Media: The Extensions of
Man (1964). Coined the terms 'hot and cold',
'the Global Village' and 'the Medium is the
Message.'*

The insatiable demand for global computer communication led to digital convergence, changing the nature of computer science and human life. The irresistible process of digital convergence leaves nothing beyond deconstruction and virtualization by software. The driverless car is a dramatic example. Virtualization will go all the way and remake law-and- order implementation in a democratic society. The *'fragmented space and time'* of the pre-electronic age assumptions are out of tune with cyberspace. The General-Purpose Computer has dragged unacceptable standards of backward compatibility into the future. In McLuhan's global village, the hot medium is software, but the cold message is *'crime pays.'*

Virtualization will proceed, but unless something changes, law enforcement and the standards of democracy will fall. It can be seen already in the echoing aftermath of the 2016 US Presidential election. The disagreements stain democracy, and more examples will only worsen things. Truth evaporates on all sides.

The medium of the electronic age is software, but the software in General-Purpose Computer Science cannot be trusted. The disastrous consequences of the contested election all started with an innocent mouse click. If it were a girder in a bridge or a building, there would be an inquisition, as there is with the Boeing 737 MAX catastrophe. The unfortunate pilots and Ms Clinton's campaign manager, John Podesta, cannot be the scapegoat. General-purpose computer science is unfit for virtualizing the life-supporting applications of a trusting democratic society. It is the crime of the age of electronics that must be stopped.

It all began when von Neumann took over from the founders of the Church-Turing Thesis. He ignored their most important discovery in his rush to leave his name in the future. His assumption of blind trust and memory sharing directly conflict with the teachings of the λ-calculus in the Church-Turing Thesis. It did not matter while in the batch processing mainframe age of standalone batch processors. However, digital convergence took off when the microprocessor emerged, and the Church-Turing Thesis became essential to the future. Not just as a nation or to solve the problems at Facebook and Google, but as a global digitally converged civilisation.

However, the virtualization of society continues with the imprudent pre-electronic age assumptions of von Neumann, timesharing, and Cold War bureaucracy as Identity-Based Access Control. Unelected strangers now dictate unacceptable rules as monopoly suppliers while foreign hackers and malicious malware pervert individuals and society in a global village that exposes democracy to corruption by enemies. These problems are human errors. The same errors that drove Charles Babbage two hundred years earlier.

Babbage's *Infallible Automation* is needed. To prevent the breakout of Artificially Intelligent Weaponised Malware, nothing less will do. Crimes against society will continue and grow in scale and success rate. A deadly, invisible dictatorial force is usurping traditional law and order. The General-Purpose Computer is outdated and a hand-me-down from a bygone pre-electronic age when the global scale of the electronic future was a mystery, technology was severely limited, and networked malware and hacking could not occur. Now, the electronic village exists, and digital corruption matters.

A violent collision occurs between civilisations and disruptive authoritarian software of General-Purpose Computer Science. The Mueller Report, an Investigation into Russian Interference in the 2016 Presidential Election and the grounding of the Boeing 737 MAX in 2019 exemplify this collision. Figure 31 summarises the transition underway and the steps needed to survive in a digitally converged, fail-safe information age. There are numerous alternative scenarios, but all are unacceptable. Some may lead to a Nuclear War, others to enslavement by branded suppliers or Artificially Intelligent Malware.

On the one hand, experts and industry claim their computers work, but on the other hand, civilians suffer. When innocent citizens die because the captain and a co-pilot lose a fight for control of a software application, the writing is on the wall. The battle takes place every day. They are increasingly impacting the whole nation. The struggle is the same. It takes place out of sight, between the human-inspired forces of good and evil that fight it out on the lawless, dictatorial digital frontier of General-Purpose Computer Science.

While experts continue to insist there is no problem if best practice procedures are followed, the citizens suffer, and innocent citizens die. The natural catastrophe goes far beyond a few examples. It repeats every time software fails to work as expected by a user. The unreasonable *'best practices'* defined by industry are created to work around an unsolved problem in General-Purpose Computer Science.

The loss of control has broad social consequences that cannot be solved by *'best practices.'* At the core, the problem is the dictatorial architecture and unfair privileges in General-Purpose Computer Science. The computer fails to meet the standards established by Charles Babbage and proven by Alonzo Church. When *Infallible Automation* does not exist, human errors, criminal corruption and ruthless dictators take over.

THE FIGHT FOR CONTROL

The fight for control over software applications cannot be dismissed as a user training problem. It is because unfair privileges distort expectations and human results. General- Purpose Computer Science is tilted against the citizen user. Experts and criminals take easy advantage to strike without warning, harming the innocent. The tilted infrastructure enslaves us all to branded industry standards that do not work but from which no one can escape.

This negative situation destroys the freedom to progress. Worse still, it converts monopoly suppliers into the tools of spies, autocrats, and criminal dictators. Authoritarian rules cannot override a citizen's rights. Freedom of action and equality do not exist in the General-Purpose Computer. Unidentified conspirators use the unfair, tilted platform for personal gain.

Circa	The Frontier of Computer Science	Period
2000 BC	Babylonian Abacus	
500 BC	Roman Abacus	
1614	Napier's Logarithmic Scale	Pre-Electronic Society
1625	Oughtred's Slide Rule	
1840	Babbage's Two Engines	
1930	Alonzo Church's universal model ofcomputation and the λ-Calculus	↓
1936	Turing's α-machine	
1945	Von Neumann's Shared Memory Architecture	

1947	Earliest Transistors	PioneeringPhase ↓
1950	Early Virtual Memory	
1960	Centralised Software Supervisors	
1965	Berkley/Multics Timesharing/IBM 360	
1969	Early ARPANET	
1970	Fabry's Magic Number Machine	
1974	Capability-Based Computers (PP250 & CAP)	
1983	Personal Computers (IBM et al.)	
1990	Early WWW Browser & public email	
2000	Serious Malware Losses	
2010	Massive Data Breach Damage	
2017	Intelligent Malware Attacks	
2019+	Dictatorial software tragedies with loss of life	
2021?	Accept trusted computers need Capabilityad-dressing as Church-Turing Machines.	The Civilisation ofCyberspace ↓
2025?	Government-enforced Church-TuringStand-ards	
2030?	Industrial Strength Global Computer Science	
2040?	Virtualised Cyber Democracy Legislation	
2050?	Point of Singularity Reached	
2075?	Life 3.0 is well Underway.	
2100++	Safe and Secure Virtualised Democracy	

FIGURE 31: MIGRATION TO THE ELECTRONIC VILLAGE OF THE 21ST CENTURY AND BEYOND

THE POINT OF SINGULARITY

The driving force behind this transition is the progress of artificially intelligent software towards the point of singularity. It is when the power of software exceeds the power of the human brain. After this formidable achievement, humanity is a second- class citizen subject to enslavement by a supernatural force. The General-Purpose Computer is the perfect tool for suppressing cyber-society through artificially intelligent weaponised malware.

The three sources of software breakout are to be resolved.

- First, software defects become bugs that go wild in the hands of unskilled citizens. They spring out by surprise, creating unchecked, unresolved worst-case conditions. The 737 MAX trouble is of this nature. Artificially Intelligent Malware will lead to more accidents as virtualised society expands.

- Second, Artificially Intelligent Malware will discover new back doors, avoiding passwords and access control lists to assume the privileges of a targeted virtual machine. Attacks will be staged, and the strike will be unstoppable when fully assembled. The well-known click-jack attack that activates Artificially Intelligent Malware will destroy a business or enterprise with a single innocent mouse click.

- Finally, user hacks misuse the unfair privileges of a branded computer. The massive data breaches that steal secrets from anywhere in cyberspace will be catastrophic when powered by Artificially Intelligent Weaponised Malware.

Nothing in General-Purpose Computer Science can solve the *'too complex'* to test problems created by increasingly complex software leading to Artificially Intelligent Weaponised Malware. Every existing application will be immediately subjugated. Without software modularity, the complexity only gets worse. Virus scanners and firewalls, like patches, only addressed known issues. Unknown, so-called *'zero-day'* attacks are always unaddressed. As software grows into complex networked applications, cybersecurity cannot remain an afterthought; it must be designed once and inherited by all as a physical, mass- produced, digital machine where cybersecurity is converged with the expression of software functionality.

Data-tight software only exists in a Church-Turing Machine. Its extreme form of Capability Limited Addressing uses object-oriented machine code to obey all the laws of the Church-Turing Thesis. The telecommunications industry pioneered it for networked cybersecurity. All rules are local and atomic by nature, and order is structured. The symbolic DNA of intrinsic directional relationships detects all sources of errors and delivers immediate justice to any infections on the spot. Its architecture remains stable, generation after generation, but it improves with each age in response to environmental changes, including the containment of artificially intelligent malware.

The electronic villagers must adopt the Church-Turing Thesis and the λ-calculus as a machine to survive an uncertain future. It is not a hopeless task, but it will take some effort and urgent government coordination.

THE WILD FRONTIER

The frontier of cyberspace is not some remote, distant location in a foreign land. It exists within every digital computer. A frontier exists where mathematics is mechanised, as the software is read by hardware and computer instructions start to execute. Its dynamic activity results in work performed by a computer. Before the electronic transistor, the transition to a digitally converged global electronic village seemed out of the question. Now, decades later, the home run of digital convergence will interconnect everything, and everything will include a computer. Everything will be driven by software; cybersecurity will be vital because trusted software is essential to the future of society.

The result of the forces unleashed remains to be fathomed. Beyond computer hardware, there is software, communications, automation, robots, AI, bugs, malware, cyberwar, and deliberately networked interference. These forces mix, and if unconstrained, they are shamelessly dangerous to a peaceful society. The only single point of control over all these forces is the instant software interacts with hardware. At this instant, the laws of science apply, and the rules are exquisitely clear.

When software interacts with hardware as a Church-Turing Machine, the laws of the λ-calculus apply. The laws turn each instruction into a mathematically bound scientific expression— an engineered function executed by digitally encapsulated Turing α-machine guided by the laws of the λ-calculus. The machine executes instructions as protected digital actions in a scientific computational frame. Functional Church Instructions apply the laws of the λ-calculus and the rules of the Church-Turing Thesis to contain each functional symbol. The interconnected mathematical expressions become typed λ-calculus variables in a directional DNA hierarchy of a globally interwoven, digitally converged software network. The software is structured as a species, each with its DNA.

The security rules of *'need-to-know'* and *'least-authority'* apply. Atomically, immutable Capability Keys connect the functional atoms through defined access rights. A networked security lattice prevents the fraud and forgery caused by binary data. All this works seamlessly across global cyberspace as the edge of the worldwide network expands. Growth occurs during the expansion process as the Internet of Things is realised. The opportunity exists to convert from General-Purpose Computer Science to the Church-Turing Machine and migrate to industrial- strength Computer Science.

The critical step is for government standards to exert pressure on movement in this direction. Industry standards must be agreed upon under government leadership to prevent a branded set of standards. The standard required must be atomic and prevent the continued use of monolithic compilations. It will take some time. However, a new round of computer chips based on the Church-Turing Thesis and the Church-Turing Machine will emerge as standards are published. The Internet of Things will use these chips to secure their family of internet devices. As functional Black-Boxes in the more extensive network, the migration to the Church-Turing Machine will occur gracefully.

As this process proceeds, the transition will occur from General-purpose Computer Science to Industrial Strength Computer Science. It will not all happen simultaneously, and there will still be more significant industrial catastrophes. However, once the standards are agreed upon, progress will accelerate in every direction. The goal is to prevent the deployment of Artificially Intelligent Weaponised Malware that can easily break out from General-Purpose Computers. The threats will be contained if this software is limited to the new computer architecture.

The negative impact on the future must be prevented for democracy to survive as expected and required. Like it or not, as Marshall McLuhan understood, *'the medium is the message'*, and if cyberspace is corrupt, society is tarnished, and democracy fails. Bad actors in Russia, North Korea, Iran, China, and a host of other enemies, criminals and hackers will use Artificially Intelligent Weaponised Malware at the first opportunity to disadvantage America and her allies. Their intent is the destruction of democracy and the American dream. The only future safe reaction is to deploy Church-Turing Machines as quickly as possible.

THE ENEMY WITHIN

International enemies are not the only ones to fear; even businesses and friends spy on one another to seek an advantage, especially in cyberspace, where both detection and punishment are missing. Cybercrime means nothing if, on the spot, it is undetected. Cybercrime is the fastest-growing industry; it offers quick paybacks, especially for those in sad economies remote from a working justice system. The attacks depress digital entropy worldwide, and without a resolution, a miserable economy will be the end state for all.

Worse still, the secret agencies in the United States, the United Kingdom, France, Germany, Israel, most industries, and others with the necessary skills are fully engaged in spying. They all use the tilted playing field of General-Purpose Computer Science to their advantage. Spying is built into their business model. It undermines trust and leads to further corruption in society.

Crime is bad enough, but the problems of software breakout and dictatorial takeover are appalling. Once the point of singularity is reached, at the onset of Life 3.0, computer science must be secure and safe in the future. The failure to do so surrenders all progress to inhuman dictators. Some think the point of singularity is when computer software outsmarts humanity within a decade, but even if it takes a generation, say by 2050, there may not be time to complete the necessary changes.

Meanwhile, digital entropy and integrity continue to decay little by little and trust justifiably evaporates as innocent citizens suffer under an illegal and unelected digital government. It is encouraged by the exceptional power and financial resources of the spy agencies who fail to speak out on the problem or invest in the extreme, fail-safe form of computer science. They like the advantage they gain from the unfair privileges in General- Purpose Computer Science, and they fail to discuss the downside risks. A fail-safe, long-term democratic solution is vital to preserve American Democracy as a government of the people, by the people and for the people.

The loss of trust and the corruption of democracy are sowing the seeds of discontent and revolution. The rabble-rousers are already building their barricades on social networks. Fake news is growing, extremists are at total volume, and false advice created by flawed assumptions is misdirecting the decisions of legislators. Deep Fakes will soon sponsor political manifestos. New, severe problems lie ahead.

Washington must step up. Several steps must take place. First, the spy agencies and DARPA must support Industrial Strength Computer Science and set new standards to match *Infallible Automation* and the Church-Turing Thesis. Without their support and financial drive, the breakout of Artificially Intelligent Weaponised Malware is guaranteed, and the replacement of a people's constitutional democracy with a digital dictatorship will take place.

DARPA already instituted a so-called CRASH[99] program circa 2010, but the objective was narrow, and CHERI was the only new computer project. It has changed nothing because it remains a backwards-looking hybrid design with rings of default privilege, a central operating system and compiled monolithic software. Industrial Strength Computer Science must be the goal, and extreme Church-

---

99   The Clean-Slate Design of Resilient, Adaptive, Secure Hosts (CRASH) program will pursue innovative research into the design of new computer systems that are highly resistant to cyber-attack, can adapt after a successful attack to continue rendering useful services, learn from previous attacks how to guard against and cope with future attacks, and can repair themselves after attacks have succeeded. Exploitable vulnerabilities originate from a handful of known sources (e.g., memory safety); they remain because of deficits in tools, languages and hardware that could address and prevent vulnerabilities at the design, implementation and execution stages. Often, making a small change in one of these stages can greatly ease the task in another. The CRASH program will encourage such cross-layer co-design and participation from researchers in any relevant area.

Turing Machines must be the architecture. A proposal was made to do this then, but it was ignored. It confirms the need for outsiders to force the mainstream to research in new directions. The pre-electronic age of WWII must be resisted for the Church-Turing Thesis and a range of Church-Turing Machines to lead us forward.

General-Purpose Computer Science has done its job but will forever remain inadequate and untrustworthy. As the endless platform for cybersociety, it tragically misses the mark. It is time to move on before there is an unrecoverable national disaster. Relationships are already on the decline as trust erodes and crime grows. Attacks and software breakouts cannot be prevented in the face of unchecked and unfair privileges. The spy agencies, the suppliers, the press, the fifth estate, individuals, and society are all fighting to redefine the 21st century in their preferred image. They wish to gain a permanent advantage through the unfair default privileges in cyberspace. The tilted platform of cyber society cannot continue. Exposing the cornerstones of democracy to constant attack and hidden corruption is a national threat. Society virtualised this way will have no happy ending. Lawmakers must set the minimum standards for *industrial-strength Computer Software,* and Washington must lead the way.

THE SECURE ELECTRONIC VILLAGE

he stable, evolving electronic village that survives the 21st century will interconnect, integrate, empower, and virtualise life securely, everywhere for everything. It starts today with the Internet of Things. It must also be an *'Internet of Church-Turing Machines.'* The secure alternative of the λ-calculus must encapsulate the tragically imperfect General-Purpose Computer. An evolving infrastructure that expands as independent software machines. These applications are protected as networked Black- Box software functions in a secure evolving network.

The overlaid network of Church-Turing Machine sets the standard for removing cybercrime and hacking, but the evolution will take time. Detecting bugs and reducing costs will speed up the change. No application will ever again be too big and too complex to test. The new object-oriented machine code will protect and operate at a new level of faithful reliability. Removing unfair privileges and using Capability Keys as the tokens of cyber power distributes ownership of confidential data to the citizens.

This new cyberspace will soon be appreciated for delivering Abe Lincoln's dream that *'government of the people, by the people, for the people, shall not perish from the earth.'* Constitutional democracy will improve as power is transferred from the monolithic dictatorships of General-Purpose Computer Science into the hands of the citizens and society.

In our conflicted, hurried world, the advantages of applying the Church-Turing Thesis to prevent crimes and stop enemies and dictatorships from winning will encourage trust in virtualised democracy. The cyberwar weapons originating from the tilted, unfair privileges of General-Purpose Computer Science will diminish and eventually be outlawed as the success of Church- Turing Machines improves and the threats of Artificially Intelligent Malware are fully appreciated.

Furthermore, since the source of any attacks is identified and reliably traced by Capability Keys, enemies and criminal syndicates will be brought to justice. Law and order will grow as crimes, lies, and misrepresentations diminish. Once again, the *'medium is the message,'* but now science is pure and equal for all, and the message changes to *'Cyber Crimes are punished.'*

Progress at the CIA, NSA, FBI, the institutions of governments, industries, societies and the halls of Congress will improve with the secure, stable evolution of trusted software as a networked machine. The unfair privileges that tilt the electronic platform away from citizens to allow malware, massive data breaches and cyberwar will end as Church-Turing Machines convert the nature of cyberspace from harsh dictators to faithful servants. It is achieved as immutable digital tokens are built into computer science using the hardware of Capability Keys and Capability-based addressing.

Only then are tangibly and transparently owned tokens used by citizens to protect their destiny and confidential data in cyberspace, establishing the foundations of digital democracy. Securing the electronic village for a life in the 21st century is vital. To migrate from corrupt dictatorships to faithful servants has everlasting monumental advantages for all subsequent generations.

Human trust will grow when crime no longer wins by default and undiscovered bugs in complex software are no longer ignored and unnoticed. It is the frontier technology of computer science that fairly, solidly, and securely fights for truth, justice, law, and order as a civilised cyber society. A level platform for democratic civilisation to grow and continuously prosper because cyber wars are always won.

THE LOSS OF CONTROL

Everyone who has experienced identity theft knows the frustration of pleading with a bank clerk to restore access to your account. Its frustration knows no bounds. Days turn into weeks of anguish and lost time on all sides. It all starts when the attack is spotted, not by the Bank or the computer, but by the user. It is wrong because the first problem is to convince the bank that there is a crime, and this is nontrivial. The bank and all other organizations are designed to deny any authority to fix such issues. They think their systems are flawless.

The dictatorial nature of the General-Purpose Computer accumulates this blind power from the computer security experts who ignore customers and the bank officers who have no computer skills or security authority to update passwords. Uniting the functional jurisdiction of two isolated departments to work together and solve the problem of individual clients is a serious challenge. All the unplanned costs and lost time are unnecessary when authority and need are aligned mathematically through λ-calculus. Most importantly, when computers detect errors, scientifically, corruption is prevented. By aligning responsibility with need using immutable Capability Keys, problems evaporate, and efficiency improves.

The multiple causes of a massive data breach at the NSA and the CIA highlight the impossible nature of security in General- Purpose Computer Science. Releasing state secrets to WikiLeaks is the tip of the security iceberg. Knowing that the CIA cannot sustain cybersecurity proves there is no chance of sustaining cybersecurity with General-

Purpose Computer Science. Blindly trusted, monolithic software grows increasingly complicated, and the endless series of patched upgrades cannot fix zero-day attacks or massive data breaches. Even the slightest unfair privilege is a long-term security issue.

The General-Purpose Computer constantly leads to a *'fight for control'* as experienced by the pilots of the 737 MAX as they flew vertically into the ground fighting with the MCAS computer. It need not be this way. It makes everyone and the spirits of Charles Babbage, Ada Lovelace, Alonzo Church, and Alan Turing cringe.

The way to avoid this fight for control in General-Purpose Computer Science is to make cybersecurity transparent and scientific. The first vital step is aligning cybersecurity with functionality using the Babylonian Golden Rule. Lock-and-key ownership of data using Capability Keys puts the control of the digital frontier in the hands of the data owners. It is how the virtualised government of the people, by the people, and for the people will be achieved.

The rights won by American independence and the Civil War are again under attack, not just by the unelected privileges of superusers but by the dictatorial nature of General-Purpose Computer Science and every enemy with a grudge or a point to make. Virtualizing society this way is playing with matches while sitting on a tinderbox. A digital firestorm that moves at the speed of light without digital barriers cannot prevent catastrophe. Consider a loss of control that destroys a community overnight like Paradise, California, in the 2018 Camp Fire[100]. The virtualization of a democratic society cannot be this uncertain, and the best effort of General-Purpose Computer Science is inadequate. Equally important, different laws exist over time and between nations; one size fits will only increase conflict and hasten domination by monopoly suppliers.

---

100   Paradise, a town in Butte County in the Sierra Nevada foothills above northeastern Sacramento Valley, CA had a population was 26,218. On November 8, 2018, the Camp Fire destroyed Paradise and the adjacent three communities. Life for these citizens will never be the same.

THE LEVEL PLAYING FIELD

The technological assumptions from the pre-electronic age tilted the playing field of General-Purpose Computer Science. These machines are designed by monopolies and for monopolies to gain commercial advantage. Any upgrades are made on their backward-compatible timeframes. It is outdated and unacceptable. The future of the nation and constitutional democracy is at stake. Law and order must remain in the hands of citizens in society to control the digital frontier of cyberspace and meet their need for data privacy and a trusted government.

The software must be media-tight, and information must belong to the data owner. The data access rights must be under lock-and-key held by users. The technology for this level of scientific playing field to serve faithfully and in full is proven. It is found with a Church-Turing Machine. Any unfair privilege in cyberspace will be seen by Artificially Intelligent Malware and used to cause harm. The abuse of power in society that drove patriots to shed their blood for freedom is a precious gift. To submit to criminal interests, enemy attacks, and dictatorial domination is tragic.

Patriots must stand and fight. Freedom, equality, and justice as a government of the people, by the people and for the people must not be surrendered to the corruption of General-Purpose Computer Science. The trusted foundation stones of society cannot be perverted by tilted platforms that favour the few over the many. Constitutional democracy is openly challenged because malware is ignored, and hackers are unnoticed. They harm society by using unfair privileges from the backwards- facing industry using WWII defaults of the standalone mainframe.

Trusted software is essential to shape an enlightened cyber civilisation. Trust is a bubble that only survives locally. A global bubble always bursts. It was the issue with General-Purpose Computers when trust in shared memory became an international concern impossible to sustain. Global trust is built from local trust. The bubble bursts because cybercrimes and hacking are undetected. Software mistakes must be detected and stopped on the job. The arm of the law must work locally to arrest the guilty *red-handed* to face immediate justice.

Furthermore, cybersecurity must protect every citizen in their data-tight cyber suit. A minimum government standard for individual privacy, such as industrial strength computer science, is vital. Computer science cannot survive on a flawed assumption from WWII, and the dependent industries, societies, and nations will all fail together.

TRUSTED GOVERNMENT

Traditionally, democracies trust their institutions to remain independent, and the laws enforce this trust. However, the virtualised General-Purpose Computer alternative is not the same. Cyberspace is full of centralised, common failures that impact machines at once. A ransomware attack that freezes 10,000 computers overnight is not the only example. The independence of the branches of government is undermined by a cyber platform that cannot be trusted. Trust is fragile and easily fractured by poor government. In the pre-electronic age, trust was palpable, law and order were physical, and justice was designed to be speedy. None of this is valid with the opaque General-Purpose Computer. Things that once worked reliably are easily broken, and even the experts are unsure why. It leaves citizens frustrated and open to gossip and fake news. Malware weaponised by Artificial Intelligence brings ever more significant threats with increased certainty.

International conflict is fought in cyberspace but spills into the streets of every major city. It is now a fight for the soul of the nation. The larger war is about freedom, equality, and justice. The spoils are once again the survival of a government of the people, by the people, for the people. Losing this fight means a government run by dictators, criminals or monopolies, and the citizens will be increasingly harassed by undetected crimes. It is a rerun of feudalism.

The virtualization of society must still be founded on freedom, equality, and justice. The power must be in the hands of the citizens. It is a level mathematical, global platform where democracy can survive, reverses the WWII assumptions made by von Neumann and removes the centralised, privileged dictatorship of time-shared operating systems. The corrupt foundation supplied by unchecked monopolies, supervised by all- powerful but unelected superusers and corrupt international interests must be vanquished.

Monopoly suppliers cannot dictate new laws, and unelected administrators outsourced to questionable foreigners cannot rule the 21st century. Outsourcing is dangerous, especially for low-cost operations in corrupt locations. It is cyber suicide. The level playing field of the universal model of computation and the laws of λ-calculus is how nations can coexist, survive, prosper and grow in the age of an electronic global village, a digitally converged cyberspace filled with Artificially Intelligent Weaponised Malware.

INDUSTRIAL STRENGTH SOFTWARE

Social stability in the 21st century requires Governments to stop spying on their citizens. There is more to gain from stable international security than spying. As established by the λ- calculus, law and order prevent spies, malware, and hacking. Allowing computers to spy enables other crimes. However, spy agencies are mighty branches of government, and their *'reaison d'etre'* is spying. Instinctively, they resist all attempts to change. It is one main reason cybercrime and hacking remain unsolved; the spy agencies use government money to pay industry to do their will. However, the war in cyberspace cannot be lost. Spy agencies must support the Church-Turing Thesis and the Church- Turing Machine. Any setback will soon be recovered in a new system of international cyber transparency.

Given that governments and industry must recognise the vital need for Industrial Strength Computer Science, the dilemmas of the electronic village will all be solved. The software stays on track when the universal computation model is installed in cyberspace's basement as a Church-Turing Machine. The computation scope is locally limited to a form that matches the function. It is how trusted security will flourish.

In this universal model of computation, Turing's single tape α-machine matches Richard Dawkins[101] *'selfish gene.'* They are the

---

101   The Selfish Gene is a 1976 book on evolution by Richard Dawkins, on Adaptation and Natural Selection (1966) where the term 'selfish gene' is a meme to express a gene-centered view of evolution.

survival machines for dynamic, living software species. Turing's α-machine is a computational digital gene, a survival machine of inherited methods for self-preservation. It includes the typed digital boundaries of Capability Limited Addressing that isolate and terminate infections on contact. Even the slightest error is an early warning of an unapproved cyber-attack. Immediately blocking and quickly killing these threads is the automatic response to terminate attacks on contact, axing the attack before any harm occurs. The Capability Keys, trace criminal activity, and speedy justice are transparent, which enhances public trust in a democratic society.

SERVING SOCIETY

$\mathbb{C}$ onverting to the Church-Turing Machines adds Capability Keys to control every software application. Moreover, the universal model of computation not only guarantees that computations follow mathematical rails with limited access rights but also keeps confidential data under user control. The λ- calculus mathematically levels the playing field by separating function abstractions from data to allow user data to remain in the owner's domain. The domain is a namespace where downloaded function abstractions perform a requested service. The λ-calculus namespace is a cyber-suit for the colonization of cyberspace.

Deterministic law originates from the λ-calculus meta- machine enforced by Capability Limited Addressing and supervised by citizens as individual data owners. In this future- safe and fail-safe platform, flawless functionality serves the user faithfully. Safety features are defaults, replacing unfair assumptions in General-Purpose Computer Science. Even AI software is mathematically constrained and prevented from breaking out. Because detection is on the spot, there is no delay between the first sign of an attack, the alarm and immediate recovery steps. As such, no data is lost, stolen or otherwise damaged.

THE DREAM MACHINE

S ecure global cyberspace is a lofty goal. When achieved, it is the ultimate machine, a gift to every future generation. The General-Purpose Computer was essential but was not a monumental, everlasting result. It hides a dangerous split personality that is a constant and unacceptable threat to society.

It is a disruptor with two characters. One is Dr. Jekyll, and the other is the killer, Edward Hyde. As Dr. Jekyll, General-Purpose Computer Science has permanently improved life in the 20th Century, but the killer Edward Hyde lurks throughout cyberspace. Hyde kills trust in a 21st-century cyber society. Hyde introduced cyberwar, cybercrime, and cyber dictatorship. Hyde has no place in the future. Its schizophrenic General-Purpose Computer will destroy the future of wealthy nations and the lives of citizens in cyber society.

The split personality exists because the Babylonian Golden Rule of the Abacus is missing. The form must follow the function for atomic security to stay in line with mathematical functionality. Without the λ-calculus, the spirit of Edward Hyde breaks out to cause intolerable trouble in an opaque, outdated contraption defined after WWII, before the microprocessor, when technology was severely limited. Things will rapidly deteriorate as Artificially Intelligent Weaponised Malware grows. The software is unpredictable because the branded conventions and unfair powers discombobulate mathematics.

Instead, the modular programs in a Church-Turing Machine match software functions to software form, keeping stored digital media data-tight, function-tight and right on target. It protects software atomically from a break-in attack by malware or hackers and break-out attempts by undiscovered bugs and dangerous A.I. software. The cellular chains of parallel threads keep instance variables private by navigating the DNA strings as fail-safe computations. At every clockwork instruction step, the mathematical symbols of the calculation are scoped, typed, bound, and defended. No freedom exists to do more or less than is expressed scientifically on a virtual chalkboard.

THE FINAL WORDS

As the biggest, most powerful machine ever built, the software in Cyberspace is also the most dangerous to human life. This software machine must be designed with far more caution than other technological innovations. The ultimate failure to apply science to the full and protect the innocent should, one day, be prosecuted and punished. The life-supporting services of virtualised democracy require Babbage's flawless degree of *Infallible Automation*. Nothing less will do.

Infallible automation calibrates software without any unchecked dark side. Software, like other technologies, must be engineered to protect life. Thus, the split personality of General- Purpose Computer Science must be outlawed by decree. The software cannot become the ultimate, most dangerous fifth column and the most severe threat to civilised progress.

Networked cyberspace turns individual *Edward Hyde* programs into Weapons of Mass Destruction. As former CIA Director Leon Panetta stated, *'It isn't beyond possibility that a Pearl Harbor-type attack can be launched against us from a laptop.'* The advent of Artificially Intelligent Software made this threat a certainty. However, the continued growth as an Internet of Things makes a transition to the Church-Turing Thesis achievable.

The Church-Turing Thesis and Church-Turing Machines prevent inevitable tragedy. By building a stable electronic platform for the survival of evolving generations of cyber civilisation, a monument to the future is achieved. The citizens tune their national monuments in cyberspace. These users develop the democratic DNA for their particular cyber society. Their nations' Namespace abstracts their customs, laws, educated progress, and elected responsibilities democratically. A Church-Turing Machine is the ultimate programmable Abacus, built for the endless future of cyber society. This computer network confirms the wisdom and foresight of 21st-century civilisation lasting beyond thousands of years.

Monolithic software that attempts to shortcut this objective fails the test of time. Only the λ-calculus levels the playing field, removing the unfair, autocratic privileges that permeate General-Purpose Computer Science. Through dynamic, symbolic binding, the universal model of computation safely and securely solves the need for data privacy and citizen control using Capability Keys to *Infallible Automation*. Industrial Strength Computer Science enables software survival in a hostile global environment. Regressive nations mixed with progressive societies are all forever under reconstruction. Its evolving future cannot be solved by statically compiled, monolithic software. It is only solved by pure mathematics, the λ-calculus and the entire scientific force of the Church-Turing Thesis.

Infallible Automation as Industrial Strength Computer Science must be the hallmark of cyberspace for the civilised achievement of a virtualised society. Self-determination, equality, and justice through law and order are all controlled by citizens of Cyberspace as privately protected users. Software democracy is subject to a local policy where power is delegated instead of dictated. Capability Keys define this power in cyberspace as, where and when needed. The power of Capability Keys can be withdrawn to prevent abuse by any member of an institution in a virtualised democracy. The complex woven digital media through which individuals, industries, cultures, societies, and nations forever interact must be edge controlled. It is a harrowing, software-defined machine that will oversee life that cannot be opaque, statically bound, and monolithic. It must be scientific, flawlessly mathematical, functional, dynamic, and evolving as life evolves, but it must remain as simple as the first lessons taught at school.

The point of singularity is the point of no return. By then, the super-human powers cannot be civilised and brought under the control of a democratic society. In the hands of cruel dictators, the breakout of Artificially Intelligent Malware will be worse than all wars ever experienced in the pre-electronic age.

The General-Purpose Computer is unaware of crimes in progress, and superusers or latent bugs cause unfettered harm. Unavoidably, monolithic software grows too complex and corrupt and is hard to evaluate with few available experts. No one can finger the error, quickly fix the flaw, or retest the recompiled build. Law and order are disconnected from crimes, and the police or administrators cannot help. Undeniably, the unelected operating systems are out of their depth, making unchecked, critical decisions, unsupervised by any respected civilian authority, remote from events, without the facts.

Justice and equality are missing, and centralised authority is bemused by rampant spying. It is impossible to appreciate how bad things will get because everyone is flying blind into the future on flawed assumptions, a corrupt digital infrastructure, and blind trust inherited from WWII. The pre-electronic age of von Neumann must be left behind to make progress using the full force of the Church-Turing Thesis and Babbage's *Infallible Automation*.

Making the challenge worse, outsourcing administrator tasks with superuser permissions to some of the lowest cost but most dangerous places on earth is both national and industrial suicide. Every email, password, and line of software is open to examination, while individuals are exposed to blackmail. General-purpose computers epitomise the *'old, fragmented space and time patterns of the pre-electric age.'* Any connection to mathematics written on a chalkboard at a university is opaque. Compilers, operating systems, and byzantine software security systems, not to mention undiscovered bugs, unseen hackers, undetected malware, and blind trust, make it impossible to check activity against intent. Nothing is guaranteed to work as expected, consistently, and forever. Worse still, monolithic software unavoidably grows, ages too quickly, and stalls progress. Industry stagnates, and citizens are hurt and die.

Instead, computer science must function flawlessly and faithfully to serve society and science. Only this avoids further catastrophe and the increasingly grim future stretching endlessly ahead. Computer software must mimic the Abacus and survive for eternity. Computers must follow the Slide Rule to encapsulate functions as clockwork forms, achieve Babbage's fail-safe standards that remove all interference through Infallible Automation, and deconstruct centralised, monolithic power to work for citizens instead of privileged experts in branded cults.

Only one Artificially Intelligent Malware breakout, one accidental but catastrophic bug or one well-targeted act of cyberwar will end the virtual world built by General-Purpose Computer Science. It will crash the nation. One such crash is too many.

Time-shared General-Purpose Computer Science cannot achieve the goal. The design must be better, the software cannot be statically compiled in advance, and binary gaps, cracks and unfair privileges must be removed. Functional software must interwork dynamically as a physical machine through the immutable Capability Keys as pure mathematics. Scientifically checked components must work in harmony with the dynamic progress of Civilisation. Service providers cannot hoard data; personal data belongs to the secure digital shadows of private citizens.

The Church-Turing Thesis and Church-Turing Machines is the electronic age solution. It is a monumental step in the Civilisation of Cyberspace. Virtualised Democracy is run by citizens using infallibly automated mathematics. A fool-proof gift like the Abacus and the Slide Rule for the good of every generation yet to come.

Then and only then, Ada's dream, her anthem for abstract and immutable truth, for intrinsic beauty, symmetry and logical completeness, will spring to eternal life—a scientific machine with a language for a global society. The Dream Machine will flawlessly address every aspect of the natural world using the λ- calculus for transparent certainty. Church-Turing Machines are scientific instruments vital to translating the principles of pure scientific functions into flawless practice for secure, open-ended Civilisation.

www.ingramcontent.com/pod-product-compliance
Lightning Source LLC
Jackson TN
JSHW012357160225
79072JS00016B/136